# HYDRAULIC TRANSPORT IN FLUMES

## BHRA Information Series

Number 2

Editor: N. G. Guy MA, MIInf. Sc.



BHRA
THE FLUID ENGINEERING CENTRE

British Library Cataloguing in Publication Data

Hydraulic transport in flumes.—(BHRA information series; no.2)
1. Hydraulic conveying 2. Flumes 3. Solids
I. Guy, N.G. II. Series
627 TJ898

ISBN 0-906085-90-X

BSDCS7

Published by BHRA, The Fluid Engineering Centre
Cranfield, Bedford, U.K.



ISBN 0 906085 90 X

# CONTENTS

# BHRA Information Series No. 2

# Hydraulic Transport in Flumes

## Introduction

This continues the occasional series of publications from BHRA which combines the resources and expertise of the Association's Research Groups and Information Department to provide "information packages" on topics of current interest within the general field of hydraulic engineering.

BHRA is an independent non-profit distributing research association, founded in 1947. The interests of its research groups cover a wide range, the main fields of study being physical and mathematical modelling of flow problems, high pressure fluid power systems, pumps and pipeline transportation of solids, and sealing technology. The Information group provides library and technical information services for research engineers both within BHRA and in its member organisations. It also organises a successful programme of international conferences as part of its information dissemination activities.

The use of pipelines as a means of transporting bulk commodities is now widely considered and well documented. However, over short distances and where gravity driven flow is possible, hydraulic transport of solids in open channels – launders or flumes – may be an attractive proposition. Little has been published specifically on such systems, although they have been used in some industries for many years.

This book brings together in one volume material previously published by BHRA in various forms on this means of solids transport. Included are a literature survey of slurry transport in flumes, particularly in the mining industry, papers presenting analyses of fluid-solid flow in flumes and optimization of flume geometry for solid transport and reviewing one particular study of hydraulic transport of solids in open channels for coal mining, and a list of selected references to supplement those used in the literature survey, with examples of patented systems. Although the specific applications included are in the mining industry, reference to a range of industries will be found in the literature survey and the selected bibliography. Similarly the analytical papers will be applicable to a wide range of transport problems.

**N.G. Guy**
**Editor**

# A LITERATURE SURVEY ON SLURRY TRANSPORT BY FLUMES

H. H. Lau

CONTENTS



BHRA Project
Project No. 13195

## SUMMARY

This survey studies the solids transport capabilities of flumes, the various design parameters, and applications appropriate to the mining industry. Implications for the pulp and paper industry are not discussed, though 2 references are listed. Literature on hydraulic swarf removal systems are also included.

Owing to the scarcity of published literature, most of the papers were obtained from the coal mining industry, though this survey is not meant for the coal industry in particular.

## 1. INTRODUCTION

With the advent of hydromining a re-awakened interest in the potential of solids transport by fluming has emerged. This is by no means a new concept in solids transport, but the possibility of low installation and minimal running costs under favourable conditions make it an attractive proposition. This review is an attempt to draw together what literature is available on the subject to act as a guide for potential users or designers of solids handling flumes.

According to the definition used by Taggart (2), "a flume (or launder), as distinguished from a chute, is a trough for transport of solid material mixed with enough water to impart definite fluidity to the mixture".

Although flume transport of solids is usually associated with alluvial mining and for transport and disposal of tailings (15, 16), it is also used for transporting coal. Indeed, flume transport of coal was used in the early 1900's in U.S.A. (7) and New Zealand (6), but its importance was not realized until the recent interest in hydraulic coal mining. Its importance can be best underlined by the fact that all new mines which are to be worked hydraulically in the U.S.S.R. or New Zealand are so designed that coal from all places in the mine can be flumed to a central point (5, 8). The coal is then raised to the surface by a high pressure hydraulic transport system, usually by pipelines. (3-6, 11, 13, 14).

Another important application of flumes is in the pulp and paper industry. However, it is not within the scope of this survey and hence it will not be mentioned in any detail. Two references (23, 24) are listed for interested readers. Reference (23) is a very comprehensive literature survey on this subject while reference (24) may also be used for other applications.

Most of the papers give only practical experiences and none have any detail theoretical consideration. A great deal of experimental work has been done on sediment flow in open channels (26, 27). Experimental work using a model flume transporting fine particles is described by Kleiman (16).

References (17-19) are solely concerned with hydraulic swarf removal.

## 2. TRANSPORTATION CRITERIA OF FLUMES

As the pressure on the open surface is always that of atmospheric pressure, the hydraulic grade line coincides with the water surface (28). This statement is well depicted in the definition sketch by Novak et al (25, Fig. 1). Since there is no pressure energy to drive the water in an open channel, the only energy causing flow is elevation energy. However, there is a limit to the elevation for effective fluming. (See section 4.3 on gradients of flumes).

Viscosity and frictional drag are two of the resistances opposing flow. Frictional drag can be reduced by suitably chosen lining materials (see section 5) and wear can be reduced by specially designed cross sections. (See section 4.2).

Figure 38 of Goncharov's book (29) depicts the mechanism of grain-detachment. The grains are lifted by the eddies and kept in suspension owing to the suspending action of eddies moving up from the bed. Hence, it is obvious that in order to transport slurry effectively, i.e. to keep all particles in suspension while being transported, the flow must be turbulent (16, 27).

## 3. MODES OF TRANSPORT

Lovell and Kindig (1) characterized the modes of transport into 2 categories: (a) non-deposit transport, during which no particles remain long in contact with the inclined surface; and (b) deposit transport, during which particles may reside upon the flume bed for appreciable times. No explanation was offerred with regards to the flow velocity.

However, Kleiman (16) in his experimental work using copper tailings observed the same phenomena and suggested that there was a transitional velocity region over which the flow changed from non-depository to depository with decreasing velocity. It is in this region that the coarse particles are not in suspension but moved by saltation along the bottom of the flume. This inevitably causes wear on the flume. Nevertheless this depository characteristic had been utilized to prevent wear while transporting floatation tailings. One mill in Miami deliberately kept the flow velocity low to allow sand to settle along the bottom and protect the flume from wear (15). However, this must be kept under control; otherwise the flume may be damaged under the cumulative load.

## 4. FACTORS AFFECTING FLOW

The velocity in an open channel may be calculated using Chezy's formula

$$V = C\sqrt{RS}$$

or Manning's formula

$$V = \frac{1}{n} R^{0.66} S^{0.5} \quad \text{(S.I. Unit)}$$

where

V = velocity of flow
C = a coefficient
R = hydraulic radius
S = slope of flume
n = Manning's coefficient of roughness

Since the capacity of the flume is a function of the flow velocity, the factors affecting flow velocity will invariably affect the capacity.

### 4.1. Roughness of Flume Material, n

Lazier (24) investigated the effect of wall roughness on the capacity of flumes and concluded that a flume material of lower 'n' value would result in:

(a) higher velocity for same grade and same hydraulic radius;

or (b) lower grade for same mean velocity and same hydraulic radius.

Hence, flatter gradients can be achieved using low-resistance linings.

### 4.2. Hydraulic Radius, R

Hydraulic radius is defined as

$$R = \frac{\text{cross-sectional area}}{\text{wetted perimeter}}$$

The shape of flumes used in coal transport is generally trapezoidal (4, 8). However, the Russians (5) use a different shape and claimed that it reduces wear on the sides of the trough section (see Fig. 2).

For a rectangular flume, the best hydraulic radius is attained by having a bottom width which is equal to twice the water depth (23). Semi-circular shaped flumes tend to become blocked due to the bridging effect.

### 4.3. Gradient of Flume, S

One of the prerequisites of flume transport is that there must be a slope in favour of the transported load (7). Reported gradients varied from as high as 58.33% to level (2)! There is a tendency for blockage to occur in steeper flumes (2, 6).

For the transport of coal, different authors quote different optimum gradients (see Table 1) but generally agree that $4^o$ is the minimum gradient for successful fluming.

| Source | Optimum Gradient |
|---|---|
| Scott (6) | $2.3^o$ |
| (4) | $6^o$ |
| Weinberg (3) | $2.3^o$ |
| Banerjee (8) | $4.1^o$ |
| Parkes (10) | $4^o$ |
| Weeks et al (13) | $5^o$ |
| (12) | $4^o$ |

Table 1. Optimum Gradient for Fluming Coal

Parkes (10) reported that flumes, together with a proprietary system called wet belt, had been used successfully on a gradient of $1^o$. It was also reported that a new transporation system had been developed which was capable of moving coal slurries on a gradient as low as half a degree.

A modified Chezy's formula as used by the Russians for calculating slurry flow velocity is

$$V = C\sqrt{R\ (S - 0.012)}$$

where

0.012 = the "coefficient of gradient" for coal-water ratios between 1:3 and 1:10 (0.007 for a glass lined flume)

Hence using this empirical formula, the minimum gradient for fluming may be as low as $0.69^o$ (1.2%) for a standard flume and may be lower with flumes fitted with low-resistance linings.

However, from (2, 10, 11, 16) there seems to be a definite relationship between the size of material transported and the minimum gradient of the flume. Figs. 10 and 11 of Simons et al (27) show a decrease in fluming velocity with increasing percentage of fine sediments in the water. Kleiman (16) did the opposite and observed that by decreasing the percentage of fine particles, there was a tendency for greater settling. Kalinske (26) stated the same phenomenon. It might be owing to this extra buoyancy of the finer particles that it is possible to flume them at a flatter gradient.

Scott (6) stated that if the elevation is greater than 1:4 (14°), there is a tendency for the coal to race ahead of the water and cause blockage or even jump right out of the flume. If steep gradients are unavoidable, stepping of flumes at suitable intervals may alleviate the problem of wear and blockage (6).

The transportation capability of flatter-gradient flumes may be improved with a copious supply of water (23). Where water is not readily available, this will increase the operational cost.

Table 2 highlights some of the points mentioned.

| Location and Source | Montana (U.S.A.) (7) | Stockton (New Zealand) (7) | Cascade Creek (N.Z.) (7) | Suanda Colliery (India) (8) | Michel Colliery (Canada) (11) | Mitsui (Japan) (11) | El Teniente (Chile) (16) | Baidajewskaja-Severnaja (U.S.S.R.) (5) | Russia (17) | Taggart (2) | |
|---|---|---|---|---|---|---|---|---|---|---|---|
| Length (km) | 2.8 | 9.2 | 11.57 | | | | 60 | 3 | (22-25) x $10^{-3}$ | | |
| Capacity | 35 - 45 ton/hr | 150 ton/hr | 8 ton/hr | 50-55 ton/shift | | | 40,000 ton/day | | | | |
| Water Requirement cu ft/sec | 1.58 | 2.5 | 3 | | Water/coal ratio=1:1 | Large | | | | | Graph of Solids Throughput vs. Water Requirement, see Weinberg (3) |
| Material Transport | Coal | Coal | Coal | Coal | Coal | Coal | Copper Tailing | Coal | Swarf | Mineral Tailings | |
| Material Size | | | | | 200mm max. | -3mm | | | -10mm | | |
| Minimum Gradient | 2.3° | 1.59° | 0.46° | | 4° | <1° | 0.172° | 4° | 0.29° | 0° | |
| Minimum Velocity | | | | 4.6ft/s | | | 1.2m/s | 5.0 to 5.6 ft/s | | | |

**TABLE 2**

### 4.4. Solid Concentration

Kleiman (16) stated that flow conditions are independent of the density of the slurry, and hence the percentage of solids has no bearing on flow velocity. Also, under gravity flow the percentage of solids does not change the flow conditions. His experiments were carried out with solid concentrations of 28% - 63% by wt. Settling occurred and results were erratic with solid concentration greater than 53% Cw.

However, Kalinske (26) propounded that the presence of sediment might decrease the mean velocity.

## 5. FLUME DESIGN AND LININGS

Flumes are generally made with very simple steel construction. They can either have flanged joints with rubber seals to prevent leakage or merely overlapping joints. Both have been proved successful (4, 8, 20). Other common materials used for making flumes are wood, concrete, and polyethylene (20, 21).

Both Keogh (20) and Newton (21) reported that flumes made of low density polyethylene are superior to mild steel and rubber for abrasion resistance. Also gradients of the flumes at Kaiser Resources were successfully reduced from $7^{o}$ to $4^{o}$ using wear-resistant low-friction plastic liners (10).

However, in New Zealand, three wooden blocks of white pine or red beech laid with the grains at right angles to the direction of flow are most commonly adopted. It is found to be more suitable than glass, rubber, and steel liners for fluming coal (8).

Other lining materials are ceramics, enamels, slag, cast iron, and concrete. Comparison of the merits of these materials was not available.

## 6. ADVANTAGES AND DISADVANTAGES OF FLUMES

The most attractive advantages of flumes are the low initial capital cost (8, 13), and that they require negligible routine maintenance (7). Safety is another inherent characteristic.

They are flexible, not only in installation (8), but also in the kinds of material which can be transported. Another advantage is that the system requires minimal or zero power.

The main handicaps are that the gradient must be in favour of the transported load (7), and there must be a plentiful supply of water.

Reference (4) contains a comparison on flumes and pipelines.

## 7. OTHER USES OF FLUMES

Lee (22) described how an improved flume design increased the efficiency of a settling tank.

Kuprin (17, 18) described how swarf may be transported over a short distance (22 - 25m) by pulsed liquid or booster jets. Nomograms (19) are available for designing hydraulic swarf removal systems.

## 8. MAIN POINTS ARISING FROM SURVEY

(i) Flow must be turbulent;

(ii) Optimum operating gradient is $4^{o}$ for transporting coal. Further reduction is possible;

(iii) Trapezoidal section is the most common shape;

(iv) Low-resistance linings can increase the capacity of the flume and reduce the angle of inclination;

(v) System is flexible;

(vi) System is safe and economical.

## 9. RECOMMENDATION FOR FURTHER WORK

It is clear from this survey that what the design engineer lacks is a cohesive set of data enabling him to optimise flume parameters for a particular application. Therefore, a series of tests is required to assess the load carrying capability of flumes by examining the interaction between the flume parameters (slope, shape of cross-section) and the solids parameters (density, size distribution, concentration).

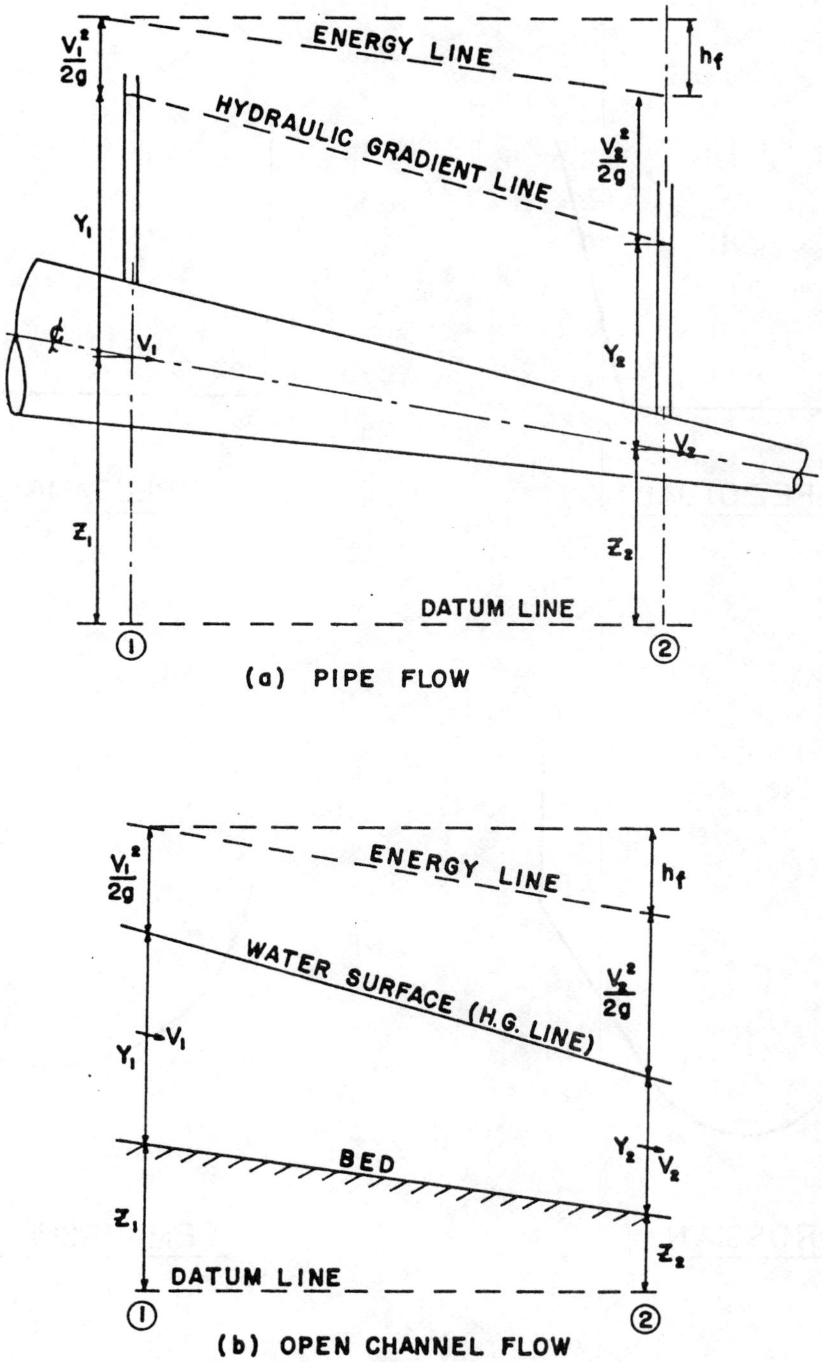


(b) OPEN CHANNEL FLOW

FIG. 1. DEFINITION SKETCH

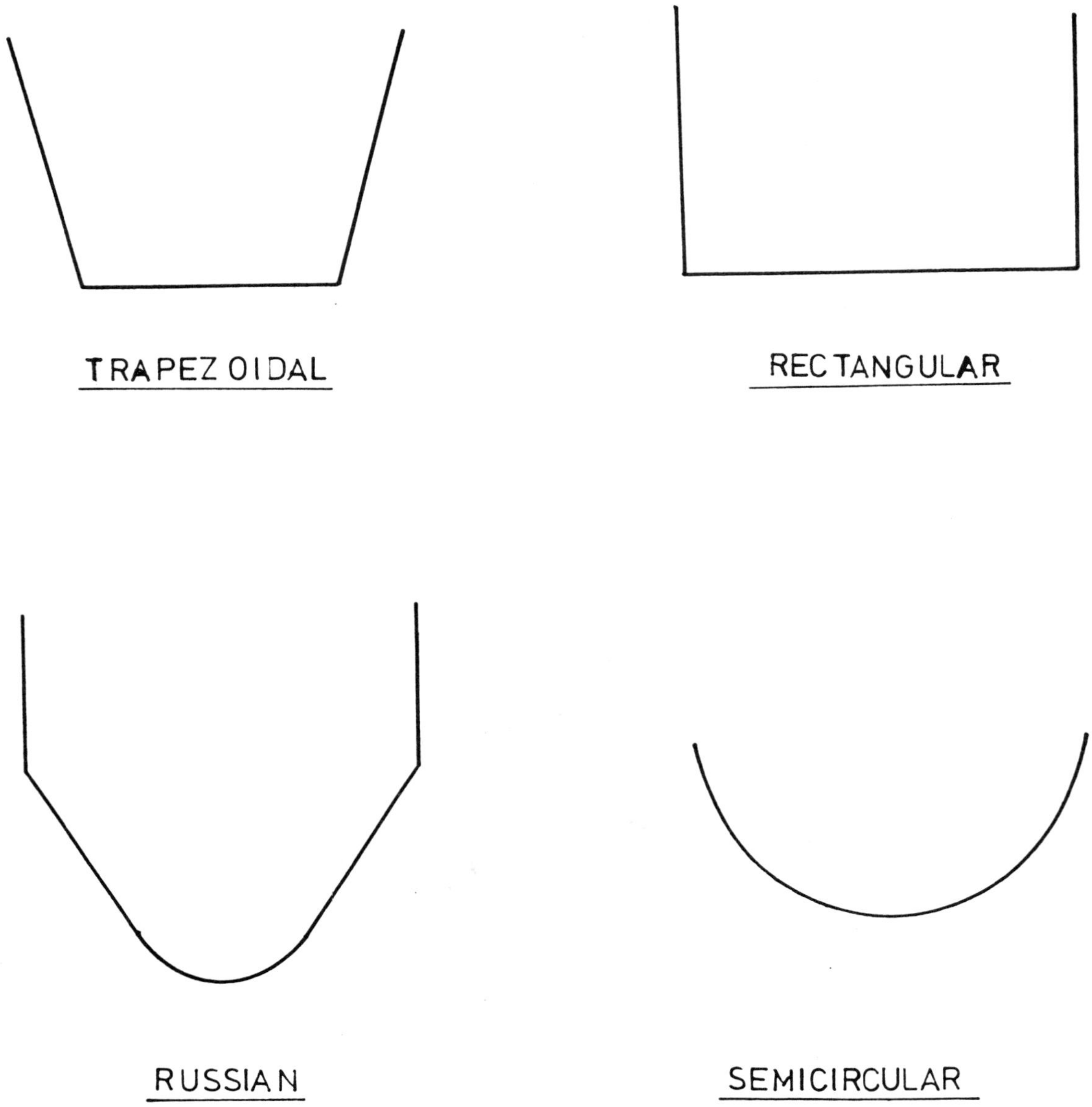


FIG. 2. EXAMPLES OF FLUME CROSS-SECTION

## 10. REFERENCES


1. Lovell, H.L. and Kindig, J.K. "Launders". In "Coal Preparation", pp. 9-85 - 9-91, edited by Leonard & Mitchell, published by American Institute of Mining, Metallurgical and Petroleum Engineers.

2. Taggart, A.F. "Launders". In "Handbook of Mineral Dressing Ores & Industrial Minerals", pp. 18-68 - 18-79. John Wiley and Sons, Inc., 1960.

3. Weinberg, S. "Hydraulic Transport of Coal". Colliery Guardian, November 8, 1962.

4. Confidential Report

5. "Hydraulic Mining in the Soviet Union". Extracts from "Visit of A German Delegation to the Soviet Union, July 1970". NCB Translation MRDE/DRJ/242

6. Scott, R.W. "Some Practical Experiences with Hydro Mining on the West Coast", Private Communication.

7. Watson, W.B. "Flume Transport of Coal", Colliery Engineering, 36, 429, pp. 474-477 (Nov. 1959)

8. Banerjee, S.P. "Hydraulic Transportation of Coal - A Review of its Application in Suanda Colliery", Jnl. Mines, Metals and Fuels, 12,7, pp. 226-31 (July 1964)

9. Parkes, D.M. and Grimley, A.W.T. "Hydraulic Mining of Coal", Mining Cong. Jnl. 61, 5, pp. 26-9 (1975)

10. Parkes, D.M. "Progress in Hydraulic Mining of Coal". From: Proc. Joint MMIJ and AIME Meeting: World Mining and Met. Technol., Denver, Sept. 3, 1976", 1, pp. 209-218. (1976)

11. Parkes, D.M. and Lindsay, J. "Hydraulic Transportation and Dewatering of Coal", Society of Mining Engineers, AIME, 262, pp. 6-10. (March 1977)

12. "Hydraulic Mining" Trade Literature of Kaiser Resources Ltd., v/o Licensintorg, Mitsui Mining Co. Ltd.

13. Weeks, P.G. and Carr, T.L. "Hydraulic Mining", Colliery Guardian Annual Review, pp. 485-92. (Aug. 1977)

14. Gontov, A.E. "Hydraulic Mining - A Refined and Growing Technological Art". World Coal, pp. 55-8. (July 1977)

15. Taggart, A.F. "Tailing Disposal". In "Handbook of Mineral Dressing Ores and Industrial Minerals." pp. 20-34 - 20-39. John Wiley & Sons, Inc.

16. Kleiman, P. "Hydraulic Transport of Copper Tailings", Proc. American Society of Civil Engineers, Jnl. Hydraulics Division. 102, HY10, pp. 1588-94. (Oct. 1976)

17. Kuprin, A.I. "Hydraulic Chip Transport by Pulsed Liquid". Machines and Tooling, Vol. XLVI, No. 12, pp. 18-19, (1975)

18. Kuprin, A.I. "Hydraulic Transport of Metal Swarf in a Trough with Booster Jets", Russ. Engng. J., 56, 4, pp. 42-45 (1976)

19. Tikhontsov, A.M. — "Nomograms for Designing Hydraulic Swarf Removal Systems", Russ. Engng. J., 54, 10, pp. 64-6. (1974)

20. Keogh, M.A. — "Plastics Launders in Mineral Processing", CIM Bull., 69, 774, pp. 93-8 (Oct. 1976)

21. Newton, F. — "Polyethylene for Abrasion Resistance", Design Engng. (Toronto), 17, 5, pp. 66-7, (1971)

22. Lee, Chesman A. — "Increasing Settling Tank Efficiency", Plant Engng., 27, pp. 126-7 (Apr. 19, 1973)

23. Benett, W.D. — "The Transportation of Pulpwood in Flumes", Pulp and Paper Research Institute of Canada, Montreal, (1956)

24. Lazier, S.S. — "The Effect of Wall Roughness on the Capacity of Flumes", Pulp and Paper Research Institute of Canada. (1958)

25. Novak, P., Nalluri, C. and Sirin, S. — "A Study into the Correlation of Sediment Motion in Pipe and Open-channel Flow". Proc. HYDROTRANSPORT 2 (2nd Int. Conf. on Hyd. Trans. of Solids in Pipes). Paper D4, organised by BHRA Fluid Engng. (Sept. 1972)

26. Kalinske, A.A. and Hsia, C.H. — "Study of Transportation of Fine Sediments by Flowing Water", University of Iowa, 1945.

27. Simons, D.B., Richardson, E.V. and Haushild, W.L. — "Some Effects of Fine Sediment on Flow Phenomena". Geological Survey Water-Supply paper 1948-G., U.S. Dept. of Interior.

28. Simon, A.L. — "Practical Hydraulics". John Wiley & Sons, Inc., (1976)

29. Goncharov, V.N. — "Dynamics of Channel Flow", Translated from Russian, Israel Program for Scientific Translations, Jerusalem, (1974)

30. Shukry, A. — "Flow Around Bends in an Open Flume", Proc. American Society of Civil Engineers. p. 713 (June, 1949)

**HYDROTRANSPORT 7**

**7th International Conference on the Hydraulic Transport of Solids in Pipes**

**SENDAI, JAPAN.** November 4-6 1980

PAPER C1

# STEADY FLUID-SOLID FLOW IN FLUMES

T. Blench, V.J. Galay,

Northwest Hydraulic Consultants Ltd., Canada

A.W. Peterson

University of Alberta, Canada

Summary

The use of open channels or flumes to transport various materials is becoming common in the mining and mineral processing industry. The transport of fairly natural sands, gravels and materials of lighter density, such as plastics and coal, has been extensively studied in flumes, primarily by engineers and geologists interested in extrapolating their findings to canals and rivers. Dimensional analysis of the flume data yields charts that clearly distinguish between phases of fluid-solid flow and permit design of flumes for transport of such materials having various densities and sizes. An example is provided for the design of a flume to transport a particular crushed coal over a long distance. Finally, comments are presented on the phenomenon of solid transport in flumes and how this relates to the wider scope of solid transport in rivers and canals.

Held at Tohoku University, Sendai, Japan.

Organised by BHRA Fluid Engineering in conjunction with the Slurry Transport Society of Japan.

## INTRODUCTION

The reason for this necessarily very condensed presentation is the authors' belief that shared observations might now benefit all three of (a) practising civil engineers concerned with natural channels that achieve equilibrium (or "regime") by building their boundaries out of their transported sediment (b) workers interested in the steady transport of river-like sediment in laboratory flumes (c) practising engineers of various specializations concerned with transporting any sediments by pipe-line, or with the flow of pure non-Newtonian fluid in pipes. Accordingly they present graphically - to avoid reliance on technicalities, theorizings and personal opinions as much as possible - the phase picture disclosed by all the quantitative data obtainable by them from sources (a) and (b), along with references to major sources. A terminating example is given in cautious use of the graphs for design or forecasting, and followed by comments on the possible relevance of total findings to pipe-line design and research.

Although the presentation is from the viewpoint of a mobile boundary - so applies directly only to pipes with active deposit - its methods and results might aid the practical analysis of clean-boundary multiphase or non-Newtonian flows.

## REGIME PARAMETERS AND EQUATIONS

The authors, as civil engineers involved professionally with rivers and offtaking unlined mobile-bedded canal systems in alluvial plains, see them as the simplest and most measurable bases for sediment transport mechanics. For, regardless of natural disturbances or uninstructed human interferences or designs such channels persist in adjusting their breadths, depths, slopes and other dimensions - see APPENDIX 1 - towards equilibrium (technically "regime") consistent with (i) the natural laws that decreed their original states and (ii) the presence of constraints. So the following quantitative narrative will start in terms of graphed numerics ("dimensionless parameters") including those that have been found to be practically constant throughout various homogeneous river or canal systems in regime. ("Homogeneity" implies that the nature and constitution of the water-sediment complex can be regarded as fixed throughout the system). Numerics allow the mimimum useful presentation to be reduced to four dimensions, and therefore to two sets of three-dimensional graphs (contoured plane sheets) reminiscent of Moody Diagrams.

Duned sand-bed irrigation canals. For duned sand-bed canals with erodible-depositable visibly smooth silty clay loam sides the classic publications of Lacey (Refs. 1,2,3) indicate that, throughout a homogenous system, the following parameters become constants: -

$$V^2/gd,\ \mu\rho V^3/b,\ gVS \qquad (1,2,3)$$

These quantities represent (i) a Froude Number in terms of a representative bank-full depth, (ii) the square of a Prandtl smooth-side shear stress, and (iii) the rate of dissipation of flow energy per unit mass of water-sediment complex per unit time. (Note: Sides were naturally eroded or deposited silty clay loam.)

Mobile-bed gravel rivers in alluvium. Observations on putatively homogeneous gravel-river systems by geologists (Refs. 4,5) and others (Ref. 6) are fewer than on sand-bed systems and do not cover such large river discharges. Dunes were believed to be absent except, perhaps, at very high stages. Constancies are accepted here (Ref. 6) as algebraically equivalent to: -

$$(V^2/gd)\,x\,(^d/D)^{1/2},\ V^5/b,\ gdS \qquad (4,5,6)$$

Of these items, (5) makes the obvious statement that sides are not smooth. Item (6), quite unlike item (3), makes shear stress constant throughout a homogeneous system. Explanation of Item (4) needs the original equations which show that, for both phases,

$$V\ \alpha\ (\frac{d}{x})^{1/4}\ \sqrt{gdS} \qquad (7)$$

where, in the Lacey sand-bed phase, $x = F_b$, and, in the gravel-bed one, $x = D$. So a common flow formula applies but the "roughness height" requires more complex description when the bed is of moving dunes.

Laboratory Flumes. In the present context laboratory flumes are viewed as short reaches of tiny rivers or canals selected at random from different systems. However they have the disadvantages of (i) vertical, rigid sides, (ii) unnaturally large and small b/d values, (iii) unnaturally uniformized or dispersed bed-material particle-size distributions, (iv) gross differences in sediment-load prophesies by competing theoretical formulas (Fig. 1 and Ref. 7). So the authors prefer to treat them, like rivers and canals, in toto. Their liaison value between the two discussed canal and river phases will be displayed below in terms of the simplified general numeric relation: -

$$V^2/gd,\ S = \text{fns}\ (\sqrt[3]{(\nu g)}\cdot D/\nu,\ d/D,\ C) \qquad (8a,b)$$

where the Vig. Number $\sqrt[3]{(\nu g)}\cdot D/\nu$, replaces the common form of Reynolds Number because of its relative constancy within a homogeneous system.

## GRAPHED TOTAL PHASE PICTURE

Flume picture geometry. Fig. 2, drawn to double-log scale, with computer aid, from a total world flume-data analysis by Cooper (8), demonstrates only the nature of the general drift of an equi-Vig (i.e. $\sqrt[3]{(\nu g)}\cdot D/\nu$ = constant) surface according to Eq. 8a; the contours are for equal C and equal d/D. The data were inadequate to define a set of surfaces - each for a small range of Vig Number - but luckily there was enough information for locating crude zones of different bed-forms. Such surfaces would have formed a sheaf rather like Fig. 3 merging, to the left, into a single sheet with straight generators sloping down at 1 upon 2 in the C = constant plane, according to item (4) above. To the right, horizontal generators would develop parallel to the d/D axis, according to $V^2/gd$ = constant.

The locus lines PO, OQ on the surface of Fig. 2, represent experiments by two observers near the state-point O. The former keeps D and q = VD constant but increases the bed-load charge C by steps; so the flow adjust to larger S values and smaller d ones. The latter runs a flume whose equipment controls C automatically so that S does not change with discharge. Here OQ might resemble a state locus as Q is increased. A two-dimensional projection of the loci of 26 different experimenters has been presented by Cooper, Peterson and Blench (Ref. (9)) to allow primary assessment of the probable weights of individual groups in forming a composite picture.

The Cooper flume plots. With Fig. 1 and the consequent need for a Fig. 3 in mind, plus a world-wide collection of flume data made and offered by Peterson, Cooper (8) attempted the cited data analysis with a view to preparing two-dimensional plots of $V^2/gd$ and S each against d/D and contoured for C; each plot would be for a small range of Vig. Number according to Eqs. 8. See Fig. 4. The data were published locally later (Cooper and Peterson, Ref. 10), have been reproduced with Cooper's thesis extracts as Appendix 2.1 of ICID (Ref. 11), and are avaialble on magnetic tape. Unfortunately 90% of the D range fell between 0.2 mm and 2 mm, D decreased consistently as d/D increased, scatter was too large to permit fitting any systematic C contours, and no scope was left for analysing the effects of extreme values of b/d. But an unexpected bonus was information that allowed the graphs to be fitted, by eye, with crude zones of different bed-forms.

Under the circumstances Cooper had to insert the equi-C curves of the single all-embracing Fig. 4 by eye, and provide a corresponding S plot; nothing more refined could be justified.

The composite picture. The preceding flume picture lies significantly in the gap between the river and canal behaviours. Thus in the diagrammatic Fig. 4 the two full horizontals to the right represent a couple of major canal systems with duned sand beds, unknown fixed C's, and D values as noted; large rivers of the same type would cover another cycle of d/D to the right. The chained horizontal marked D = 50 mm concerns a few special observations on a duned gravel river of some 350 cumecs high discharge. To the left a firm band, of slope minus 1/2, encloses a set of relatively-small, unduned gravel river observations with D values from a few millimeters up to about 300 mm; scatter in this band showed no dependence on D. Special observations on some of the rivers by Kellerhals (12) suggested to the authors the firm roughly central line of minus 1/2 inclination which is carried over, to merge with the probable D = 50 mm line at the right, by means of two alternative imagined dotted lines of which one dips and rises. These dotted lines express the authors' opinion that there are no present grounds for prophesying the nature of a phase transition. The picture for S resembles that for $V^2/gd$.

The authors feel that these two pictures' utility to river engineers parallels that of the text-book Moody Diagram to engineers aware of its various real transitions (Ref. 13 Fig. 4.2; Ref. 14, Fig. 2), the non-conformities in the large Reynolds Number zone, and the dangers of extrapolation to supercritical flow which is outside its scope.

The QF curves. The acquisition of further flume data (Peterson and Howells, Ref. 15), combined with plotting experience, suggested to Peterson (Ref. 16(a)) that a transformation to coordinates giving fairly linear equi-Vig plots would allow rapid computer analysis for system not previously obvious. Accordingly he decided to plot not $V^2/gd = q^2/gd^3$ but: -

$$QF = q/(g(s-1)D^3)^{1/2} = (d/D)^{1.5} \times q/(g(s-1)d^3)^{1/2} \qquad (9)$$

and, in the process, allow explicitly for s = 2.65 by replacing the co-ordinates: -

$$d/D,\ V^2/g(s-1)d,\ S,\ D \text{ alone},\ C \qquad (10)$$

by: -

$$d/D,\ V^2/g(s-1)d,\ 1.65S/(s-1),\ \left(\frac{s-1}{1.65}\right)^{1/3}\cdot D,\ 2.65C/s \qquad (11)$$

where, for physical reasons, the last item deliberately implies a change from concentration by weight to concentration by volume, improves fit of data points appreciably, but might raise some theoretical objections. Basic treatments of numerics are in Ref. 17-19 from different viewpoints. To demonstrate the effectiveness of these coordinates plots of low density points were made on the diagrams both with and without allowance for s.

The result was ten dimensionless linear plots, each for a partial Vig Number range, and each containing equi-C lines to avoid the extra charts. Fig. 5, from Peterson, Ref. 16, is a recently relabelled sample for rapid use when s = 2.65. It multiplies C, S and Vig. Number by co-efficients that make the effect of specific gravity explicit but reduce to unity when s = 2.65. (In the original plots the effect was tested and allowed for although the coordinates did not show this.)

Application to Coal Specific Gravity. Obviously the application of Fig. 5 to comparable flume set-ups is simple, so a sand-transport experiment could prophesy the behaviour of a comparable "coal-sand" occurrence passably. The authors prefer to exemplify from a supposed scaling of a river by a coal model because that reminds of the possible professional consequences of physically unjustified extrapolations. Thus, with properly graded coal-like material of density 1.33 and D = .34 mm for the river model, we select the Chart (Fig. 5) with the proper "size-factor" - Vig Number written to remind of g (crit. depth) equals $q^2$ - which tells us that, using dashed letters for the river prototype, $(.33)^{1/3} \times D = (1.65)^{1/3}.D^1$. That is, the model material size must be about 1.7 times that of the river. Then the captions on the left of the chart show that $S'/S = 1.65/0.33 = 5$ and $C'/C = 2.65/1.33 = 2$. For most rivers d/D would be outside of the Figure and extrapolation would be risky without field experience.

As commercially crushed coal, or other material, might have grain shapes and size distributions radically different from those of active river-bed sand the scaling formulas, above, might be unacceptably inaccurate if applied to a sand-bed model flume intended to simulate the quantitatives of a large commerical one. The formulas consider only bed-load.

The dual civil engineering approach to sediment transport. Because of the demonstrated unity and dynamical sense of the overall steady-transport picture available to civil engineers - despite the different or mixed phases that present themselves to differently situated observers - the authors now outline their assessment of the combined different civil engineering approaches. They 'see' the field-based development of fluid-solid flow mechanics as following the lines of the quantitative physical sciences (Ref. 20) via (i) prolonged measures of interrelated factors in a phenomenon, and (ii) continued searched for fitting formulas within acceptable limits of statistical significance - whose simplicity of mathematical form suggests the operation of physical laws. The irrigation engineers of the Indian sub-continent and, separately, the U.S. Geological Survey which takes care of routine river measurements with mixed engineering and geologic staff, spent several decades working on such lines before the main structure of this paper's picture could emerge. Neither party found the field measurement of bed-load to be either practical or useful (compare with the unmeasurable "roughness height" of a pipe). However, both parties were aware of the general natures of bed-load and suspended-load, and had interest in laboratory experiments.

The institutional development, via laboratory flume experiments, has resulted in a post-graduate topic "Sediment Transport" which, like civil engineering Fluid Mechanics, cannot have the rigour of its purely deductive parent Hydrodynamics. There had to be "plausible assumptions" that make some deductions and extrapolations dangerous, as illustrated by Fig. 4. Obviously practical problems should be tackled in terms of both attacks. As it would be presumptuous to rate the mass of contributions from institutional developments the authors merely offer a short list of references from which they all feel significant benefit in factual knowledge and outlook. It is (i) Gilbert (Ref. 21) on flume experiments, (ii) Bagnold (Refs. 22-24) on desert sand-dunes and laboratory and philosophic sequels, (iii) Meyer-Peter (Ref. 25) on flume experiments with unnatural grain distributions, and on the separation of rigid side effect, (iv) Einstein (Ref. 26) on (a) extension of Meyer-Peter's separation method to grain and dune roughness components and (b) the vertical distribution of suspended load,(v) Guy, Simons and Richardson (Ref. 27) on bed-forms in mobile-bedded flumes.

Relevance of River Mechanics to Pipe-Line Transport. Advice or criticism of any kind would be appreciated on how flumes might be adapted to aid pipe-line transport forecasting. The authors' possibly naive ideas include (i) comparison of flume bed-load transports of natural river-bed sands, coal mixes prepared to be closely similar, and similar mixes of other densities, (ii) repetition with selected unnatural mixes aimed to assess the effects of suspensions on bed-load transport, bed pick-up and bed drop-out, (iii) studies related to bed-forms, (iv) flume shape modifications by a "roof" and angle fillets to approximate the effect of a circular pipe with the aim of assessing the effect of an exposed pipe boundary during pick-up and drop-out (v) optical examination of turbulence structure.

Conclusion. The authors felt impelled to present the preceding exposition of river and canal hydraulics because of (i) the simple natural textures and size distribution of active-channel bed-material, (ii) the very similiar material of desert sand-dunes, (iii) the simple bed-form types, (iv) the simple dynamically suggestive conservation-type expressions for the two main phases bounding an inexpressible transition in the laboratory-flume zone, (iv) the adaptability of the formulas to allow for different solid specific gravities, (v) the background provided for future factually-directed research with different nature of transported solids in bed-movement, suspension or both, and possibly with non-Newtonian fluids describable by two counterparts of viscosity, (vi) the considerable success of river models from the beginning of the century when run by agencies combining field and laboratory knowledge, (vii) the recent commercial publication of co-ordinated field observations on, and thence of the means to avoid scour failures of bridges in alluvium (Ref. 23), (viii) the interest of Environmental Protection Agencies in the long-term consequences of uninstructed interferences with river regime, (ix) the absence from University undergraduate civil engineering curricula of River Mechanics as a partner of the now well-established applied geoscience of Soil Mechanics. Thence it seemed that pipe-lining, with its considerable field data, significant suspended loads of materials of various textures and types, sometimes non-Newtonian fluids, drop-out and pick-up conditions - but relatively small discharge range - could be a natural partner of river engineering and fluviology. So the authors would be glad to attempt answers to questions on this matter as well as on the technical side of the text.

REFERENCES

1. Lacey, G., 1930, Stable channels in alluvium. Proc. Instn. Civ. Engs., Vol. 229, Paper 4736, pp. 259-384.

2. Lacey, G., 1934, Uniform flow in alluvial rivers and canals, Proc. Instn. Civ. Engs., Vol. 237, Paper 4893, pp. 421-544.

3. Lacey, G., 1947, A general theory of flow in alluvium. Journal Instn. Civ. Engs., Paper 5518, pp. 16-47.

4. Leopold, L.B. and Maddock, T., Jr., 1953, The hydraulic geometry of stream channels and some physiographic implications. USGS Prof. Paper 252, pp. 57.

5. Leopold, L.B. and Emmett, W.W., 1976, Bed-load measurements, East Fork River, Wyoming. Proc. Nat'l Academy of Sciences, Vol. 73, No. 4, pp. 1000-1004.

6. Peterson, A.W. and Blench, T., 1980. Graphed equilibrium parameters of channels formed in sediment. Canadian Journ. of Civil Engineering Vol. 7, No. 1, pp. 93-104.

7. ASCE, Sedimentation Engineering, 1975, Manuals and Reports on Engineering Practice - No. 54, pp. 221, pp. 229-230 and pp. 629-627. ASCE, 345 East 47th St., New York, N.Y. 10017.

8. Cooper, R.H., 1970, A study of bed material transport based on the analysis of flume experiments. Ph.D. Thesis, Department of Civil Engineering, University of Alberta, Edmonton, Alberta, Canada, p. 212.

9. Cooper, R.H., Peterson, A.W. and Blench, T., 1972, Critical review of sediment transport experiments. Journal Hyd. Dn. ASCE, Vol. 98, HY5, Proc. Paper 8873, (Closure in HY 7, July 1973), pp. 827-843.

10. Cooper, R.H. and Peterson, A.W., 1970, A review of data from sediment transport experiments. Report HY-1969-ST2, Department of Civil Engineering, University of Civil Engineering, University of Alberta, and National Technical Information Center, Springfield, Va., U.S.A. Publication No. 187892, pp. 401.

11. ICID, 1974, Flume data and regime basics. Secy-General, International Commission on Irrigation and Drainage, New Delhi, India, pp. 401.

12. Kellerhals, R., 1967, Stable channels with gravel-paved beds. Jour. Waterways and Harbours Div. ASCE, Vol. 93, WW1, Proc. Paper 5091, pp. 63-84.

13. Blench, T., 1969, Mobile-bed fluviology. University of Alberta Press, Edmonton, Canada, Fig. 9.1, Para. 12.16. Photo Plates 24, Figures 30, pp. 168.

14. Barr, D.J.H., New forms of equations for the correlation of pipe resistance data. Proc. Instn. Civil Engrs. Part 2, 1972, 53, September, 383-39- and ibid 1973, 55, September, 795-810.

15. Peterson, A.W. and Howells, R.F., 1973, A compendium of solids transport data for mobile boundary channels. Environment Canada, Inland Waters Directorate, Ottawa, 1973, pp. 115.

16. (a) Peterson, A.W., 1975, Universal flow diagrams for mobile boundary channels. Canadian Journal Civ. Engineering, Vol. 2, No. 4, pp. 549-557. National Research Council, Ottawa.

16. (b) Peterson, A.W., 1978. A note regarding scale selections for mobile-boundary river models. Proc. Transport Process River Modelling Workshop. National Water Research Institute, Burlington, Ont. Canada. 1978.

REFERENCES (con't)

17. Blench, T., 1969, Dimensional analysis and dynamical similarity for hydraulic engineers. University of Alberta Bookstore, pp. 149.

18. Yalin, M.S., 1965, Similarity in sediment transport by currents. H.M. Stationery Office, London, pp. 24.

19. Yalin, M.S., Mechanics of sediment transport. Pergamon Press Inc., New York, U.S.A., 1972, pp. 290.

20. Ashford, Theodore, A., 1967. The physical sciences. Holt, Rinehart and Winston, Inc., New York and London. pp. 736.

21. Gilbert, G.K., 1914, The transport of debris by running water. United States Geological Survey. Professional Paper 86. pp. 263.

22. Bagnold, R.A., 1924, The physics of blown sand and desert dunes. Methuen, London, England, p. 265 (1960 reprint).

23. __________ 1966. An approach to the sediment transport problem from general physics. United States Geological Survey, Professional Paper 422-1, p. 37.

24. __________ 1973. The nature of saltation and of "bed load" transport in water. Proceedings, Royal Society of London, Series A, Vol. 332, pp. 473-504.

25. Meyer-Peter, E. and Muller, R., 1948. Formulas for bed-load transport. Proceedings, 2nd Meeting of International Assoc. for Hydraulic Structures Research, pp. 39-64.

26. Shen, Hsieh W., 1972, Sedimentation (Einstein). 105 pages Appendix. Available Shen, P.O. Box 606, Ft. Collins, Colorado, U.S.A. 80521.

27. Guy, H.P., Simons, D.B. and Richardson, E.V., 1966, Summary of alluvial channel data from flume experiments. US.G.S. Prof. Paper 462-I. Reproduced as App. 1, 2 of ICID (above).

28. Neill, C.R., 1973. Guide to bridge hydraulics, Toronto University Press, pp. 191.

29. IAHR, 1973, General report. Technical Session No. 6, River Bed Form, pp. 1-12. Proceedings, International Assoc. for Hydraulic Research Symposium on River Mechanics. Bangkok, Thailand, pp.9-12.

## APPENDIX 1

### FORMAL STATEMENT OF CASE FOR RIVERS (ADAPTED FROM REF. 29)

An essential of any continuing scientific qualitative analysis of data is seen to be, from the earliest stages, a periodically revised Statement of Case in the form "When, in a certain general situation, such and such quantities are specified, then, in general, the following other quantities will assume or fluctuate about definite values".

The following is suggested as a minimum for rivers including their artificial counterparts such as canals and models: -

"The time-average values of the geometric properties of average-steady self-formed rivers in sediment, namely:

1. Breadth, depth and slope,
2. Bed-form length and amplitude,
3. Meander wavelength and amplitude,
4. Surface boil period, magnitude and spacing,

are each in general, functions of the measures of:

a. Fluid discharge, density and viscosity;

b. Discharge and nature of the transported material, and nature of those boundary parts (usually the sides) that are eroded and are deposited without appreciable effect on transport;
c. Gravity."

The statement expresses each of the ten geometric quantities of items 1-4, that most fluviologists seem to visualize readily as "dependents", as a separate function of all the items of a-c. In item b, most of the numerous "nature of" terms are either poorly measurable or not measurable.

Till recently canal and river data have not justified reduction of this statement to equations more refined than the seven independent ones:

$$b, d, S, M_L, M_B, \lambda, \alpha = \text{fns } (Q)$$

(see Notations). These are applicable to the analysis of reasonably homogeneous systems of channels with similar hydrographs forms. Recent improvements in techniques of observation aim to expand the right hand side to fns (Q, D, C) but inclusion of total sediment gradation factors seems remote.

## NOMENCLATURE

| | |
|---|---|
| A | = cross-sectional area of flow = bd |
| b | = a measure of breadth such that bd = A |
| C | = bed-load charge as load per unit water-weight discharge (parts per 100,000) |
| d | = a measure of depth such that bd = A |
| D | = median bed-particle size by weight |
| fn | = function of |
| $F_b$,Fs | = bed and side factors defined practically as $V^2/d$ and $V^3/b$ |
| g | = body force of gravity per unit mass |
| q | = measure of discharge intensity such that bq = Q |
| Q | = volumetric water discharge |
| QF | = the Q-factor = $q/\left(g(s-1)D^3\right)^{1/2}$ |
| s | = $\rho_s/\rho$ |
| S | = slope measured along the flow |
| V | = space-time mean velocity across a station = Q/A |
| x,y,z | = indices |
| $\lambda,\alpha$ | = mean bed-form length and amplitude |
| $\rho$ | = mass density of fluid |
| $\rho s$ | = mass density of bed-load material |
| $\mu$ | = dynamic viscosity |
| $\nu$ | = kinematic viscosity |
| $\tau$ | = shear stress |

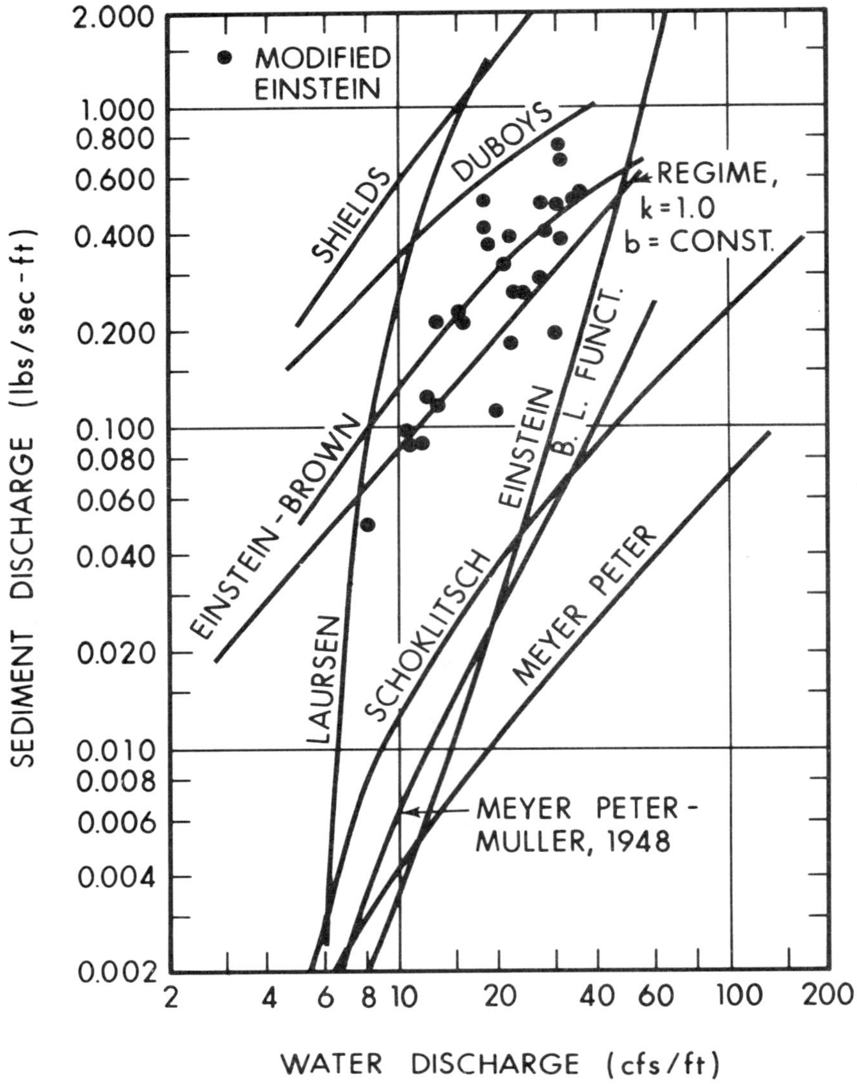


Fig. 1 Bed-load formulas compared (after Vanoni et al. Ref. 7)

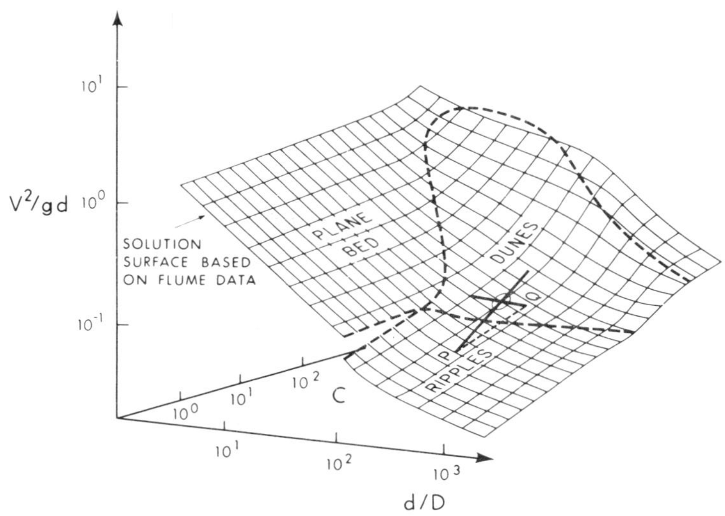


Fig. 2. Approx. Vig surface of $V^2/gd$ from flumes (after Cooper, Ref. 8)

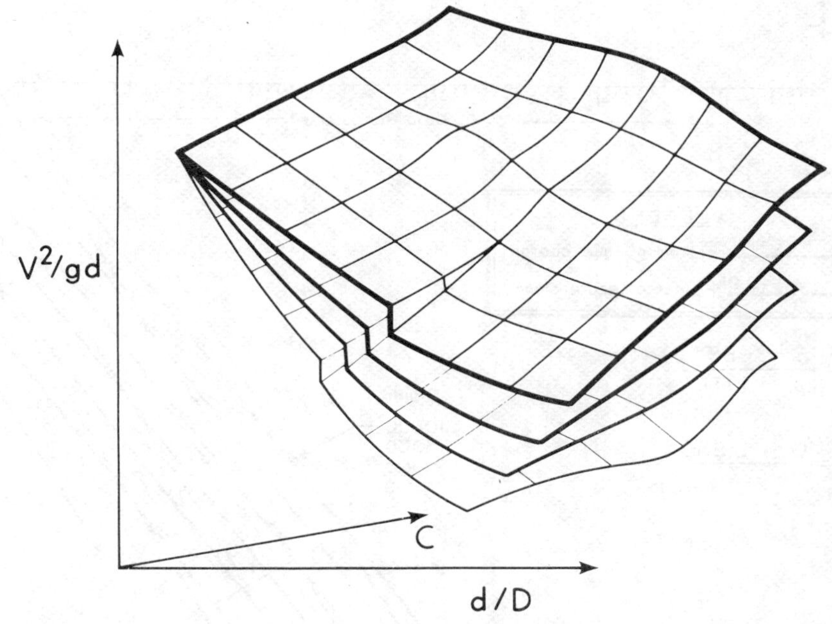


Fig. 3 Schematic Vig surface sheet.

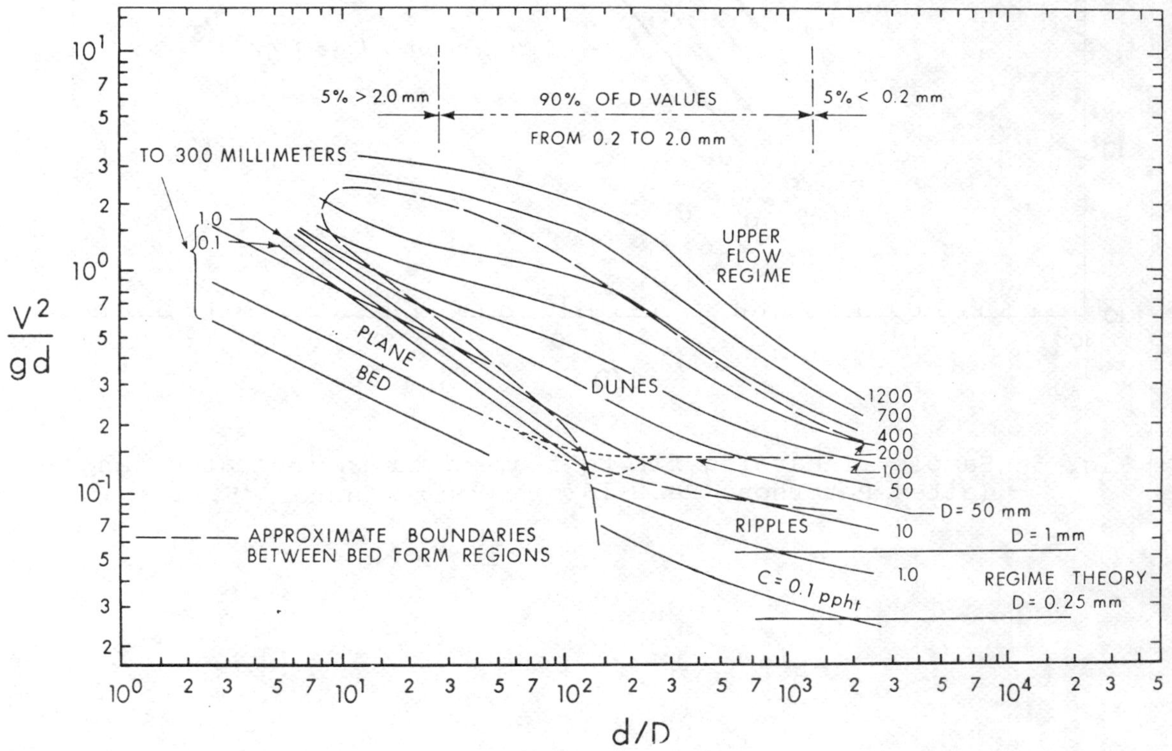


Fig. 4 Skeleton of river, canal and flume phases (after Peterson, and Blench, Ref. 6).

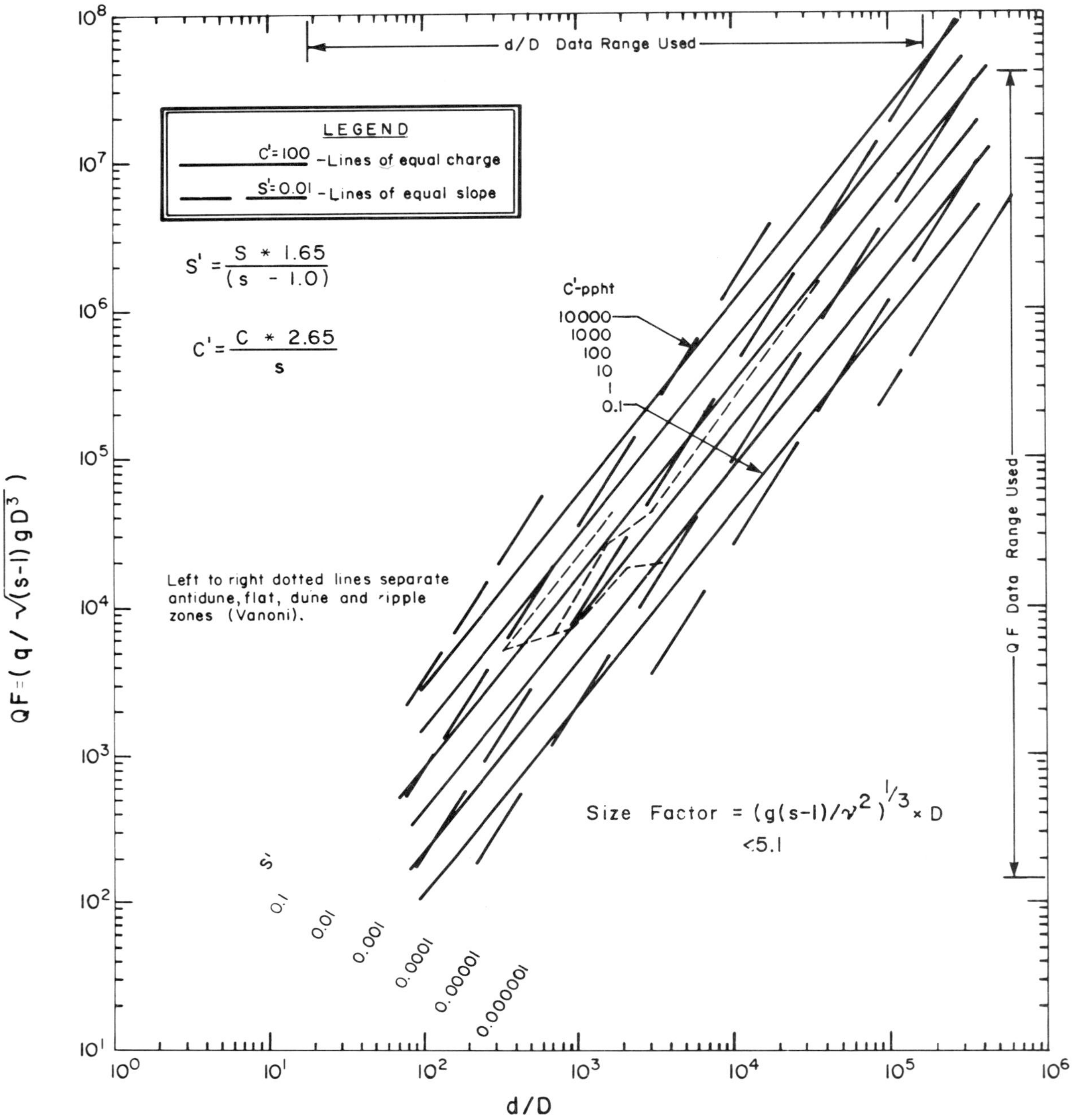


Fig. 5 Sample linear-log fit to flume data in indicated range (after Peterson, Ref. 16(b)). Size factor <5.1.

# HYDROTRANSPORT 7

7th International Conference on the Hydraulic Transport of Solids in Pipes

SENDAI, JAPAN. November 4-6 1980

PAPER C2

## OPTIMIZATION OF FLUME GEOMETRY FOR OPEN CHANNEL TRANSPORT

P.A. Wood,

IEA Coal Research, U.K.

Summary

In the dynamic equilibrium state of natural channel transport the channel cross-sectional geometry varies according to flow regime and sediment load to maintain optimum efficiency as a sediment transporting mechanism. This paper is concerned with the optimization of flume cross-sectional geometry for the efficient transportation of solid particles by hydrotransport in an artificial open channel. The use of the DuBoys equation is discussed and the importance of various parameters outlined. Controllable parameters include flow depth, gradient, particle size of solids, density of transporting medium flume geometry. Relationships are presented for various flume cross-sectional geometries from which the shear stress available for solids transport can be determined for different flow depths or flow cross-sectional areas. It is suggested that these relationships can be used for: (i) optimization of flume geometry for any given variable parameter; (ii) optimization of flow depth and/or particle size for any given flume geometry.

Held at Tohoku University, Sendai, Japan.

Organised by BHRA Fluid Engineering in conjunction with the Slurry Transport Society of Japan.

## INTRODUCTION

Rivers are natural open channels conveying both water and a sediment load, and can be considered as flume hydrotransport systems. Natural channel transport is in a state of dynamic equilibrium in which the various parameters are inter-dependent and variable. Of major importance to an artificial flume hydrotransport system is the fact that channel cross-sectional geometry varies according to flow regime and sediment load to maintain optimum efficiency as a sediment transporting mechanism (e.g. Ref. 1).

This paper is a theoretical study concerned with maximising the shear stress available for the transportation of solid particles by hydrotransport in an artificial open channel by optimization of flume cross-sectional geometry. Such optimization is desirable in order to conserve water and reduce energy requirements.

## DETERMINATION OF SHEAR STRESS

Flow in natural open channels occurs under the influence of gravity, and the use of gravity flow for hydrotransport in artificial flumes would reduce energy costs involved in pumping. In such circumstances the determination of shear stress should not involve pumping related parameters.

The determination of the ability of natural open channels to transport sediment has been approached from several viewpoints, many of which are applicable to artificial channels. Of the possible methods, this ability has often been considered in terms of shear stress which can be expressed by a modified DuBoys equation:

$$\tau = \rho RS \times 9.80665$$

where $\tau$ is the mean shear stress available for sediment transport, $\rho$ is the density of the transporting medium (often denoted as $\gamma$ ), R is the hydraulic radius (quotient of cross-sectional area of flow/wetted perimeter) and S is the slope of the energy gradient (usually taken as the slope of the water surface, but occasionally taken as the channel gradient). The constant, 9.80665, is necessary to convert the units of $\tau$ to pascals ($N/m^2$) as the DuBoys equation would otherwise give $\tau$ in units of $kg/m^2$.

Various modifications of the DuBoys equation have also been used where R has been approximated by a variety of parameters related to depth of flow, with different relationships for various channel shapes (Refs. 2,3 & 4).

The use of shear stress determinations for sediment transport studies in natural channels however, has been criticized on a number of grounds (Ref. 5). Nevertheless, the method remains a practicable one that is readily used. Shear stress not only indicates the forces available to maintain sediment transport, but provides information of the ability to initiate motion in stationary particles, and of the maximum size of particles able to be transported (Ref. 6 & 7).

## DISCUSSION OF PARAMETERS

The various parameters involved in open channel hydrotransport are interrelated and relevant ones will briefly be discussed.

Shear stress. In order to maintain an efficient transporting system and to avoid sedimentation, the shear stress should be kept as constant as possible in time and space. Shear stress should also be as high as possible to maximise transportation capacity. High values of shear stress not only enable larger quantities of solids to be transported, but also permit larger particles to be moved. As indicated by the DuBoys equation, shear stress is determined by the density of transporting medium, the energy gradient and the hydraulic radius. Certain circumstances may require higher shear stress values than the constant level maintained during normal operation of a hydrotransport system. Such higher values can be used for the re-initiation of movement in settled particles following an enforced shut-down period. Particles at rest require these higher shear stress values to initiate motion (compared to those necessary to sustain motion) as lift forces are required and particle interaction has to be overcome.

Density of transporting medium. Slurry transport systems generally use water ($\rho$ = 1000 kg/m$^3$) as the transporting medium, the density of which can be greatly changed by large quantities of suspended or colloidal particles, or additives. Other slurry transport systems, often used for fuel slurries, may use various oils as the transporting medium, while still others may have stabilizers added. Such a use of an oil medium or stabilizers will again influence the density. Liquid carbon dioxide has also been considered as a transporting medium for coal slurries (Ref. 8). However, density of the transporting medium can be pre-determined, and it is desirable that there should be no temporal or spatial variation during operation.

Energy gradient. The energy gradient of an open channel hydrotransport system can also be pre-determined and, once the system is constructed, can be assumed to be constant in time. However, unlike hydrotransport in pipelines, it is important that the energy gradient is constant along the length of the flume and so prevent sedimentation of solids, overtopping or backwater effects. The energy gradient may be determined by the environment in which the system is to be installed.

Hydraulic radius. The hydraulic radius is determined as the cross-sectional area of flow divided by the wetted perimeter. This relationship is controlled by the depth of flow and flume geometry. Hydraulic radius increases as flow depth increases and high values result in high shear stress.

Depth of flow. For hydraulic transport in an open channel this parameter should again be constant in time and space. The depth is one of the factors controlling the hydraulic radius, and thus the shear stress, which should be kept constant to avoid sedimentation of transported solids. Depth of flow is again able to be controlled, and is dependent on the volume of slurry introduced into the system.

Grain size. The particle size of solids being transported can be variable, but the maximum size able to be moved will depend on the shear stress available. For a pre-determined shear stress the maximum size of particles being transported can be controlled by a feeder-breaker or by screening. Conversely, if a certain size of material has to be transported the required shear stress must be available.

Flume geometry. Flume geometry is a factor controlling the hydraulic radius, and thus the shear stress available for particle transport. Changes in shear stress resulting from a given variation in flow are directly related to the shape of the flume. Optimization of flume geometry to maximise the hydraulic radius, and thus the shear stress, will increase the efficiency of a flume having other controlled parameters.

A variety of flume shapes are possible, the most common being an open rectangular shape, a semi-circular shape or a V-shape. Low walled flumes (low height to width ratio) are probably more practical than high wall flumes. For example, certain operational problems have been encountered when using high walled flumes in Chinese coal mines where deformation of the flume has resulted from lateral pressures (Ref. 9). Site characteristics and installation requirements may have some influence on selected flume geometry.

## RESULTS

The relationships between hydraulic radius and cross-sectional area of flow, and between hydraulic radius and depth of flow have been determined for three types of open flume geometry; V-shaped flumes (Figs. 1 & 2); rectangular shaped flumes (Figs. 3 & 4); and semi-circular shaped flumes (Figs. 5 & 6). As would be expected, all three flume geometries considered indicate an increase in hydraulic radius with increasing cross-sectional area of flow, or depth of flow.

V-shaped flumes. For a given area of flow in V-shaped flumes, Fig. 1 indicates that R is at a maximum with $90^{\circ}$ V-shaped flumes and there is a reduction in these values as flume angle becomes more acute or obtuse. For a given depth of flow, Fig. 2 indicates that higher values of R are obtained with wider-angled flumes. However, with wide-angled flumes an increased depth of flow requires a large increase in the cross-sectional area, and thus volume, of flow. In practical applications, a $90^{\circ}$ V-shape would provide maximum values of R, would not require as great a volume or depth of flow to obtain a given R as would wider angled flumes, and is less likely than acute angled flumes to trap sediment.

Rectangular flumes. The width of rectangular flumes is important in determining R for a given area of flow. With reduced flows, higher values of R are obtained with narrower flumes (Fig. 3). However, with narrow flumes the rate of increase of R with flow area is less, and at higher flows, wider flumes produce higher values of R. Individual curves for particular flume widths take on a flattened reverse S-shape with more accentuated shapes for wider flumes. The lower rate of increase in R with area of flow towards the top of the curve is associated with overdeepening of the flume so that an impracticably high depth/width ratio results. The rate of increase in R also decreases with increasing depth of flow (Fig. 4), again as high depth/width ratios occur.

Semi-circular flumes. Fig. 5 indicates that for a given area of flow, smaller diameter semi-circular flumes produce higher values of R. However, the larger diameter flumes are capable of producing the highest values of R if the area of flow is sufficient. Any particular flume diameter will have a maximum R which occurs when the flume is full. Unlike V-shaped or rectangular flumes, the depth of a semi-circular flume is finite. For given depths of flow in semi-circular flumes, higher R values are again obtained with wider flumes (Fig. 6). For individual curves the rate of increase in R with depth falls as the flume becomes full. Fig. 7 indicates the relationship between flume diameter and the maximum value of R possible (when the flume is full).

## DISCUSSION

As flume geometry is just one of the variable parameters that determine the shear stress developed in a flume, no definite relationship can be presented between geometry and shear stress. However, as discussed earlier, there is a relationship between geometry and hydraulic radius. Accordingly, it is then possible to present a relationship between hydraulic radius and shear stress if the other variables are taken into account.

Fig. 8 presents an example relationship for a given density but variable energy gradient, while Fig. 9 presents an example relationship for a given energy gradient but variable density. Such relationships can be constructed for any desired energy gradient of flume or density of transporting medium.

Figs. 8 and 9 also present transcribed particle size data (from Ref. 6) indicating the approximate maximum size of material able to be transported by a given shear stress. Actual maximum size transported however, is also a function of other parameters, including channel roughness, specific weight of particles, shape of particles etc.

If the required energy gradient or shear stress of an open channel hydrotransport system is known, it is suggested that R can be used to determine flume size and geometry. For example, if the energy gradient of a proposed hydrotransport system is 0.010, the transporting medium to be used is water (1000 $kg/m^3$) and the maximum size of particles to be transported is 100 mm, then a shear stress in the vicinity of 22 $N/m^2$ is necessary, requiring an hydraulic radius of approximately 0.22 (Figs. 8 or 9 being suitable for this particular example). Such an hydraulic radius can be obtained by a variety of flumes:

(i ) $90^o$ V-shaped flume with a flow area of approximately 0.38 $m^2$ (from Fig. 1)

(ii ) 1 m wide rectangular flume with a flow area of approximately 0.40 $m^2$ (from Fig. 3)

(iii) 1 m diameter semi-circular flume with a flow area of approximately 0.31 $m^2$ (from Fig. 5)

Conversely, for an existing flume, it would be possible to optimise flow depth and cross-sectional area, or particle size. For example an existing rectangular flume of 1 m width and 1 m depth is capable of attaining an hydraulic radius of 0.33 (from Fig. 4). If the energy gradient is 0.015 and if the transporting medium is water (1000 $kg/m^3$) this flume is capable of developing a shear stress of 46 $N/m^2$ and of transporting particles of up to approximately 150-160 mm (from Fig. 8). However, if the flume is only required to transport particles of up to 100 mm then a shear stress of only 22 $N/m^2$ is necessary (from Fig. 8). This can be obtained with an hydraulic radius of 0.15 (from Fig. 8) which only requires a depth of flow of approximately 0.22 (from Fig. 4).

These relationships determined here are theoretical, and refinements would be possible, or an empirical relationship determined.

## CONCLUSION

It is suggested that the hydraulic radius, R, of an open-channel hydrotransport system can be used for:

(i ) optimization of flume geometry and size for any other given variable parameter;

(ii) optimization of flow depth and/or particle size for any given flume geometry.

## REFERENCES


1. Leopold, L.B. and Maddock, T. : "The hydraulic geometry of stream channels and some physiographic implications". U.S. Geological Survey Professional Paper, No. 252, 54 pp. (1953)

2. Ogrosky, H.O. and Mockus, V. : "Hydrology of agricultural lands". In: "Handbook of Applied Hydrology" (ed. V.T. Chow), section 21, McGraw-Hill (1964)

3. Bradley, W.C., Fahnestock, R.K. and Rowekamp, E.T. : "Coarse sediment transport by flood flows on Knick River, Alaska". Bull. Geological Society of America, 83, pp. 1261-1284 (1972)

4. Wood, P.A. : "Competence of Hope River, Jamaica : A reappraisal of R as a function of D". Water Resources Research, 13, 6 pp. 1013-1014 (December 1977)

5. Yang, C.T. : "Incipient motion and sediment transport". Proc, ASCE, Journal of Hydraulics Division, 99, HY10 pp. 1679-1704 (October 1973)

6. Baker, V.R. and Ritter, D.F. : "Competence of rivers to transport coarse sediment bedload material". Bull. Geological Society of America, 86, pp. 975-978 (July 1975)

7. Wood, P.A. : "Sediment transport in the Hope River, Jamaica: a tropical drainage basin characterized by seasonal flow". Proc. of Paris Symposium, "Erosion and Solid Matter Transport in Inland Waters", Int. Assoc. Hydrol. Sci., Publ. No. 122, pp. 149-156 (1977)

8. Anon. : "Coal slurried with liquid $CO_2$ for transport reduces environmental impact and costs less". Energy Research Reports, 6, 6 pp. 1-3 (March 1980)

9. Lin, Yi-Lin : "Developing hydraulic coal mining technology in China today". World Coal, 5, 10 pp. 44-45 (October 1979)

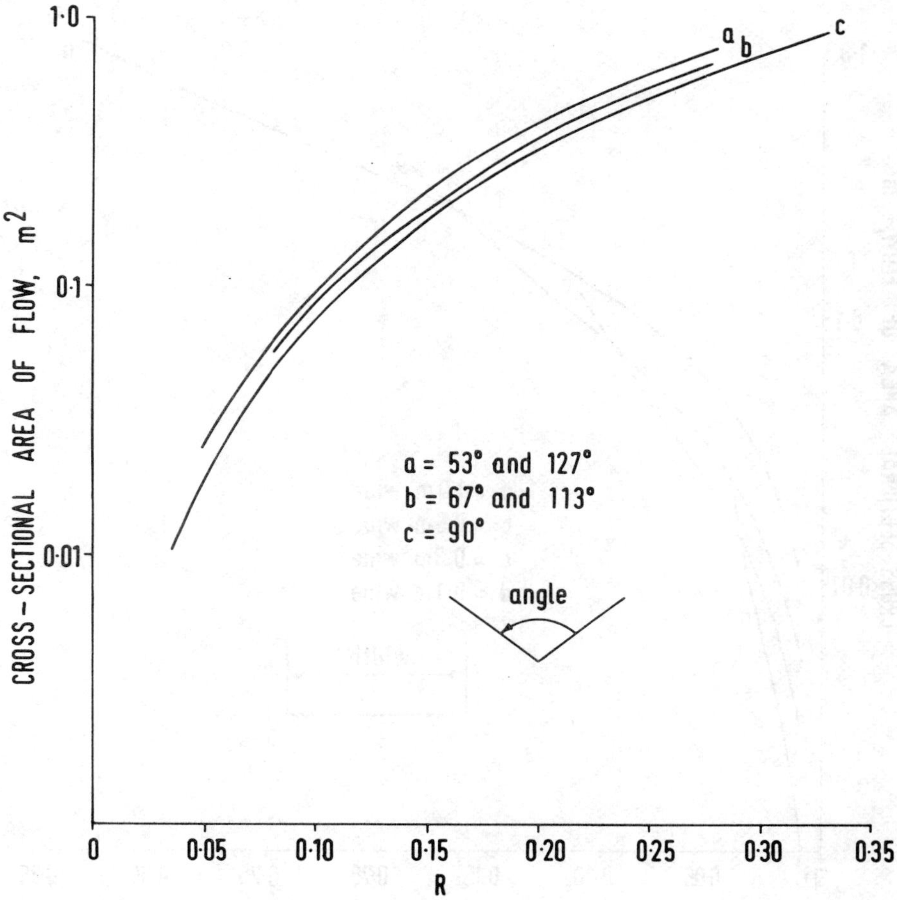


Fig. 1. Relationship between hydraulic radius and area of flow for V-shaped flumes

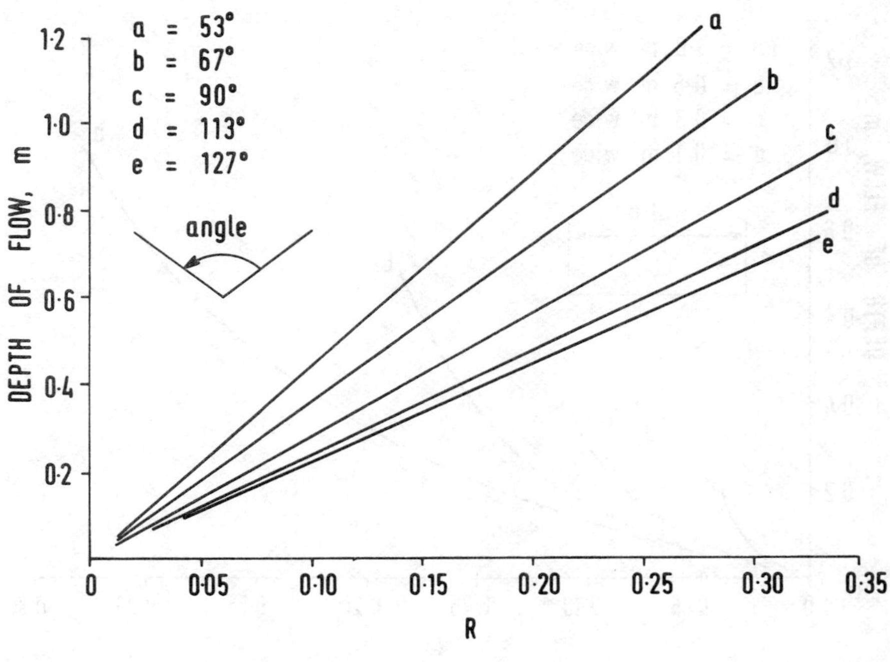


Fig. 2. Relationship between hydraulic radius and depth of flow for V-shaped flumes

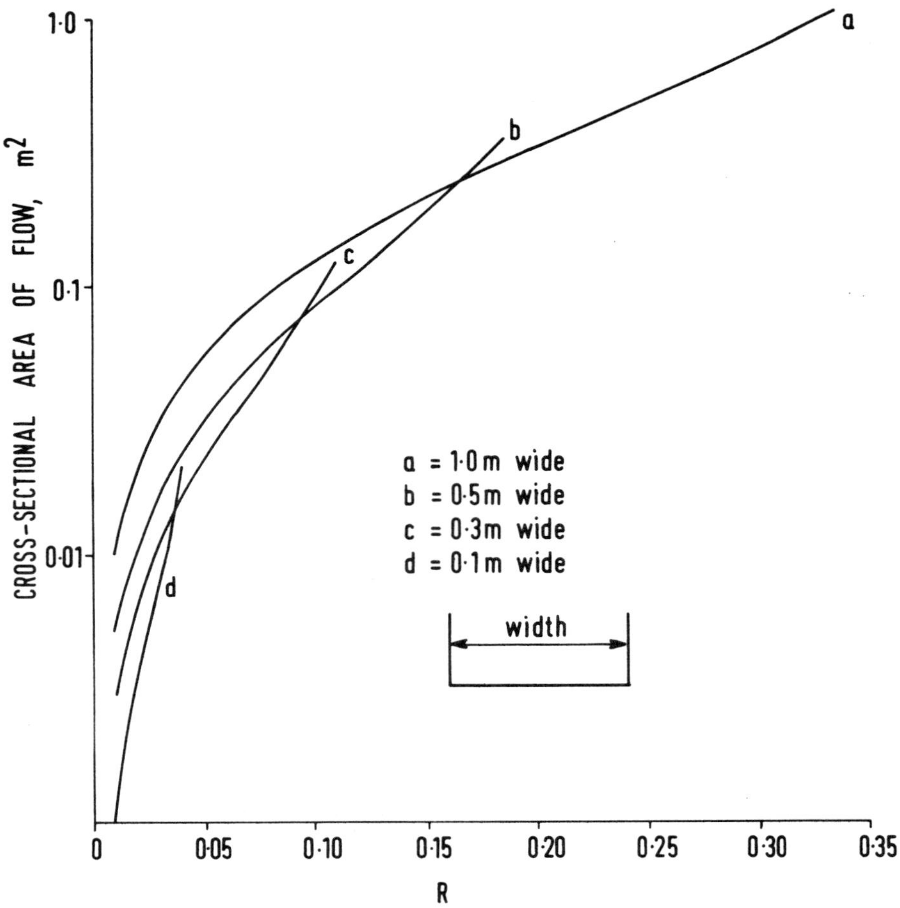


Fig. 3. Relationship between hydraulic radius and area of flow for rectangular flumes

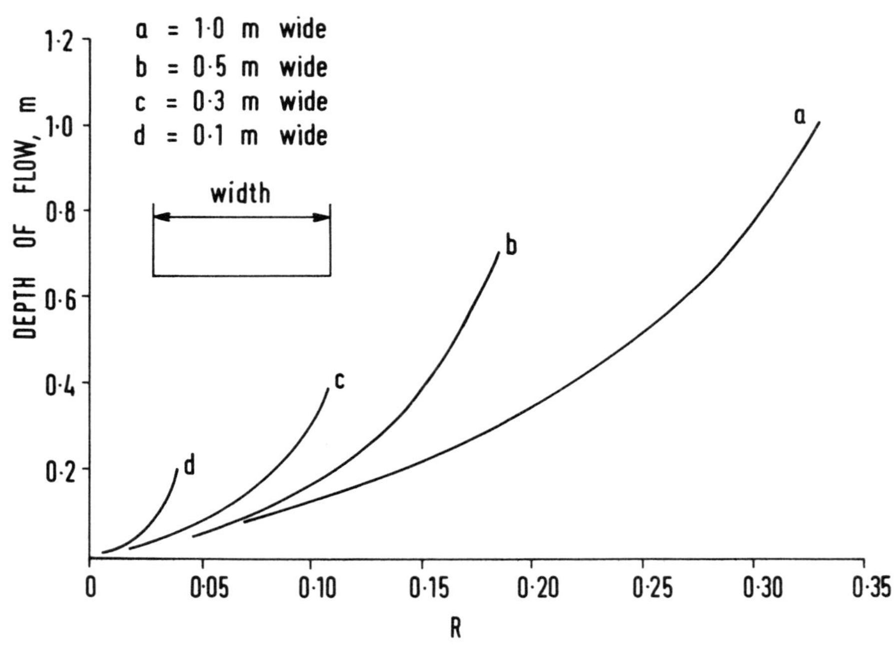


Fig. 4. Relationship between hydraulic radius and depth of flow for rectangular flumes

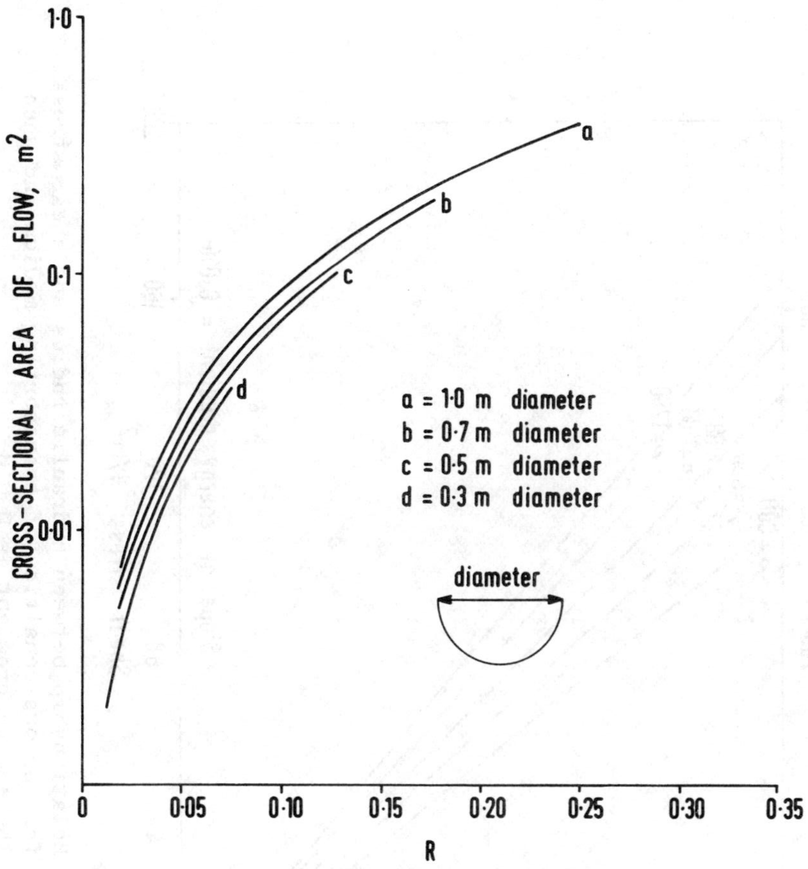


Fig. 5.

Relationship between hydraulic radius and area of flow for semi-circular flumes

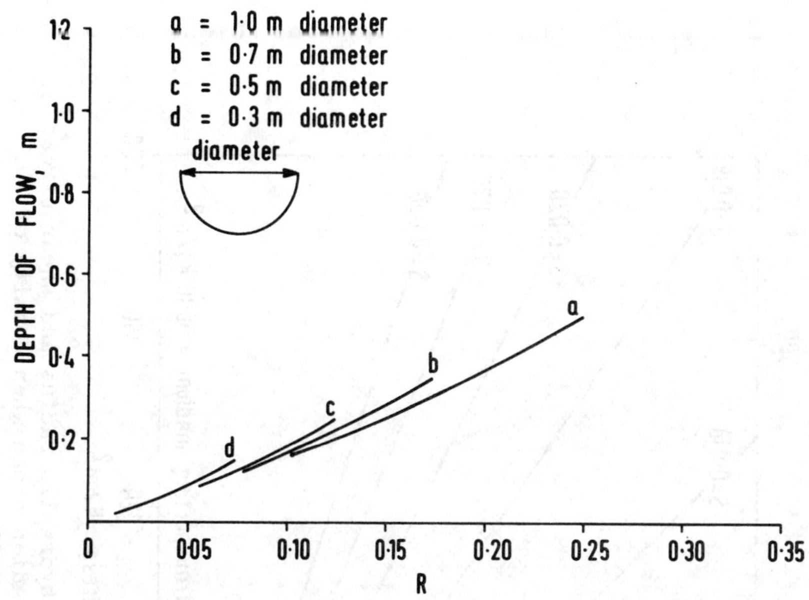


Fig. 6.

Relationship between hydraulic radius and depth of flow for semi-circular flumes

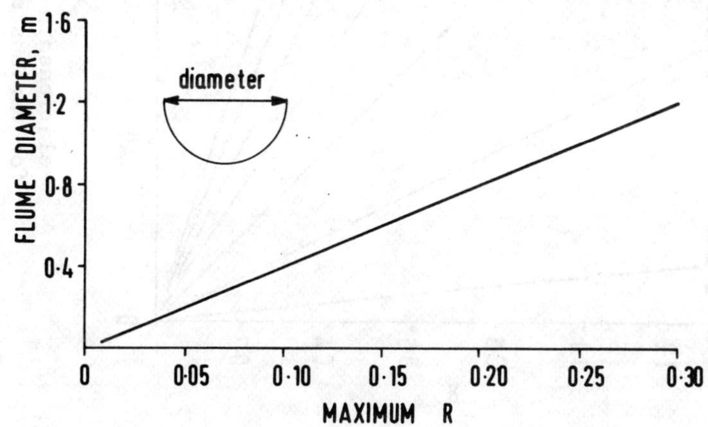


Fig. 7.

Maximum hydraulic radii attainable by semi-circular flumes of various diameter

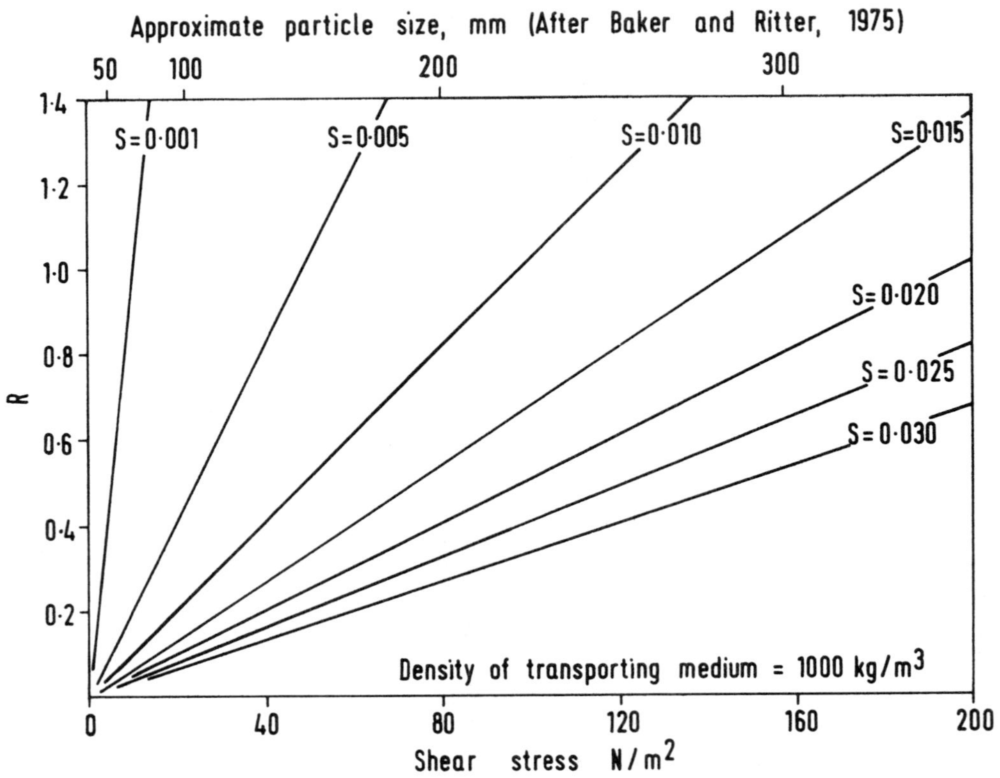


Fig. 8. Relationship between hydraulic radius and shear stress for various energy gradients and when density of transporting medium is 1000 kg/m$^3$

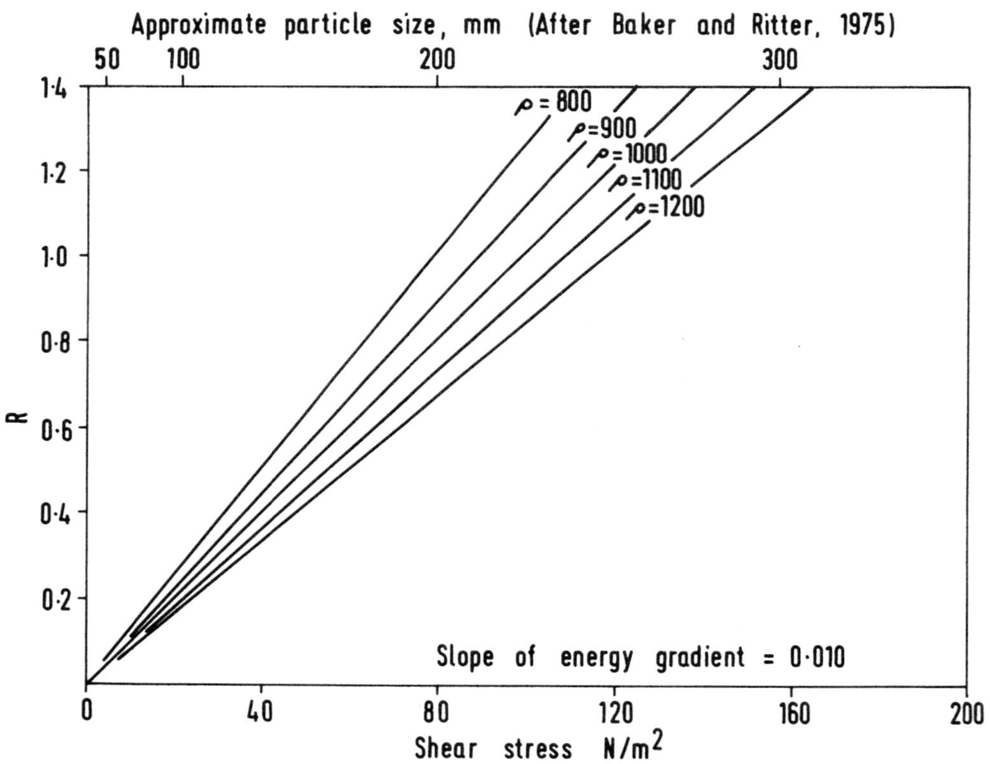


Fig. 9. Relationship between hydraulic radius and shear stress for various densities of transporting medium and when the energy gradient is 0.010

**HYDROTRANSPORT 7**
7th International Conference on the
Hydraulic Transport of Solids in Pipes

**SENDAI, JAPAN.** November 4-6 1980

**PAPER C3**

# HYDRAULIC TRANSPORT OF SOLIDS IN FLUMES IN THE MINING INDUSTRY

M. Kuhn

## Summary

Within the scope of a project of R & D activities at the Hansa Hydro-Mine in Dortmund, extensive analyses were necessary to test the characteristics of hydraulic transportation of coarse material (run of mine coal) in flumes from the workings to the pumping stations. Different designs of flumes and various materials were tested at a constant slope of 6 degrees as well as at larger inclines up to 60 degrees. For the changes of direction or special transitions various designs were tested to determine their operating characteristics and feasibility.

The influence of the angle of inclination for smaller gradients e.g. less than 6 degrees only could be determined precisely in a surface test rig. Extensive analyses were carried out with several models of flumes and coatings with transportation quantities of 200 t/h and slurry flows up to 700 $m^3/h$. For all the special designs of flumes, coatings and the various inclinations tested, the velocity, the flow structure of liquid and solids and the settling limit were determined.

Held at Tohoku University, Sendai, Japan.

Organised by BHRA Fluid Engineering in conjunction with the Slurry Transport Society of Japan.

In Hydro-mining the transportation of the extracted run of mine coal coming from the workings with an average part of 20 percent rocks suitably is effected by open flumes. The moving power is generated by gravity and the quantity of the water used for the high pressure extraction of the coal. Such flume systems may be widely ramified with lengths up to 10 000m (1).

In a physical view the reliable transportation of a definite quantity of solids needs a corresponding quantity of water and a sufficient gradient. The friction factor 'f' of the flume-material is closely connected with the angle 'a' in which the flumes are installed. Figure 1 also shows that the type of transported material influences the velocity of transportation. In slight inclination the water runs with a higher speed than the solids.

From experience data of flume transportation (2) may be recommended for the volumetric flow rate $Q = m^3/s$:

| | |
|---|---|
| Velocity of transportation | $V = 6$ m/s |
| Cross sectional area of flume | $A = 3 \times \frac{Q}{V} m^2$ |
| Ratio of solids : carrier fluid | = 1 : 3 |
| Particle diameter | = 0,25m |
| Incline | $= 6^{\circ}$ |

Besides these physical facts with regard to the economical working of the mine there are some more points of view to consider (figure 2):

- Flumes in the workings for once-only use shall have a slight weight but a good resistance against mechanical stress. The wall thickness will be dimensioned to the expected volume to be carried, so that no repair will be necessary.
- Collecting flumes in the riser headings have to resist all influences during the whole working life. Therefore replacable wear liners are suitable.
- Flume transitions shall be constructed in a manner, that the coal will not be broken unnecessarily. They are designed to be built into all geological conditions. The wear liners are replacable and the flume transitions can be used again at another place.
- Flumes in main roads have to stand over a long time with a large slurry throughput. Therefore corresponding construction or special coatings are necessary.

In hydromechanic workings with an incline below $40^{\circ}$ (figure 3) the gate roads are driven in the dip. The transportation of the extracted coal demands flumes that are lowered into the bed of the retreat road so that the solid-water mixture can reach them coming from the extraction area.

On the way from the hydromechanic workings to the shaft or a feed station (3) the slurry runs through several types of flumes (figure 4):

- The flumes in the gate roads have a small, conical form. During the road driving they are installed so as to overlap.
- In the riser headings we are still testing flumes with rectangular and semi-round cross section connected by flange-joints. In headings with a greater dip the flumes are covered by plates. Expecting a large slurry throughput at the bottom they are armoured with wear liners.
- In the main-roads only trapezoidal flumes with plastic coatings on the inside can be recommended.

At the passage from a steep flume line into a main-road with an inclination of $6^{o}$ and changing the direction, special flume transitions are used (figure 5).

The straight and conical pieces of this construction with circular cross section may be connected together in a way so that their adjustment to all geological conditions is possible. Since the cross section is always round, the slurry finds the best and trouble free way by itself. It is easy to change the wear liners. The dismantling of these flume transitions and their rebuilding at the point of the next usage is easy.

The new development for special cases is shown in figure 6. It is a steel-wire armoured rubber hose. We are testing this underground to get operational experiences of wear, working life and operating characteristics.

With all these special types of flumes - from the workings to the feed stations - at Hansa Hydromine we have gained a great deal of experience.

In the gate roads of the strong inclined formation three types of simple flumes are tested (figure 7). They all have a cross section of $0,1m^2$.

Although the trapezoidal flumes 'A' with a wall thickness of 2mm have the smallest wear, the roof pressure that has broken the prop is able to compress them in a way that is shown by '$A_1$'. The result is a blockage.

With the angular flume 'B' we also were not successful. If roof pressure happened it would be compressed like '$B_1$' or put into a position such that one flank of it lay in a horizontal direction and the slurry flowed into the road where the solids kept lying.

The semi-round flume stood the test the best in this kind of road. It cannot be compressed easily because it rises upwards. As this causes more wear, the wall thickness has to be chosen somewhat greater (in our case 4 - 5 mm).

After the extraction of the whole pillar the specific costs of the flumes in comparable gates are:

| | |
|---|---|
| Trapezoidal flume | A = 0,65 DM/t |
| Angular flume | B = 0,52 DM/t |
| Semi-round flume | C = 0,74 DM/t |

Due to the different positions of the coal seams with regard to the riser headings the connection between the flume in the gate road and the collection flume normally requires complicated adaptors. For this case an universal transition was constructed so that a manifold can be used (figure 8).

On the photo one can see the flume at the bottom normally lying in the riser heading. At the necessary point the upper part will be put on. In this case the flume of the gate road is coming from the left side. If it is turned about $180^{o}$ it also can be used for the entrance of a flume from the right side. The bounce walls in the inside of the box are installed in such a manner that they are laying in direction of the trajectory and the pieces of coal will not be further reduced in size.

In an underground main road with gradient of $6^{o}$ in a normal line various kinds of wear liners were tested to get their life time respective to the attainable throughput. The first results are shown overleaf.

| Type | Throughput | | Wear |
|---|---|---|---|
| Wood - oak | 33 000 t | | 100 percent |
| Wood - bongossi | 33 000 t | | 80 percent |
| Concrete - reinforced with Ferrodur | 45 000 t | ferrodur | 100 percent |
| Concrete - (normal) | 45 000 t | 5 mm rinsed grooves | |
| Steel flumes with wear liners | | | |
| out of St 50 . 2 | 230 000 t | | 2,4mm |
| Semi manax | 230 000 t | | 1,5mm |
| Oxar 320 S | 230 000 t | | 1,3mm |
| St 52. 3 | 230 000 t | | 1,3mm |
| St. 70 | 230 000 t | | 1,2mm |
| V 2 A | 230 000 t | | not yet perceptible |
| Plastic RCH 1000 | 560 000 t | | not yet perceptible |

The steel flumes with wear liners were weighed before they were installed. The first measurements of wear were made after six months use underground. More accurate results will be given after the closing down of the mine and the dismantling of the flumes. The same will be done with the plastic wear liners that are working in another main road.

For getting some more knowledge about the influence of the gradient of the flumes extensive analyses has been carried out at a test plant on the surface (figure 9). Most of the aggregates of the equipment showed in the presentation of Prof. Geller were suitable for our studies. Only the construction for the installation of the flumes had to be built with the equipment for changing the gradient of the flume line over the whole length.

Four types of flumes were tested with a constant volumetric flow rate to find out at what incline and what quantity of rocks a blockage happens.

The time for using this test plant was very short, so we were only able to test the overlapping flumes with the high turbulent flow at the over-lapping joints. The flumes connected by flange-joints could not be fabricated during this time.

In figure 11 are shown the results of these four test series. All these flume types have the same cross section, two of them are coated with plastic. For all tests we only used rocks because the rock content of the run of mine coal is 20 percent in average with peaks up to nearly 100 percent. Each type was tested with a flow rate of 560 $m^3/h$. Beginning with a fixed gradient of the installed flume the quantity of the rocks was raised from 50 t/h up to 200 t/h till the next blockage happened. After that the same flume was tested with the next gradient. So we got the bounderies of blockages of the four types of flumes. In the diagram it can be seen very clearly the advantage of the plastic coating. Carrying 200 t/h rocks the trapezoidal flume coated with plastic works reliably at an incline of $4^o$. In the semi-round flume without coating we of course have the first blockage. In the underground installations without any turbulence at the flange-joints will be somewhat better still.

To gain the raficiation for these results a new test flume line was installed underground with which a gradient from 7 down to 2 degrees is adjustable. The trapezoidal flumes are flanged - not overlapped - and armed with plastic wear liners. Beginning with an angled of 7°, each month the gradient was decreased. At the time of writing one of these lines has reached 4.5° and, with the normal throughput, no blockage has occurred.

Very interesting results were also gained through the measurements of the transportation speed. (Figure 12) The velocity of the water was measured with small pieces of white coloured cork, and the velocity of the material with coloured rocks. The velocity of the mixture is calculated.

The digram definitely shows the influence of the dip in regard to the velocity of the slurry. What is not seen at first is that the slip between the material and the fluid gets smaller at a higher dip due to the larger velocity. But this is calculated exactly.

In the example shown in this diagram, at an incline of 2.5° and the velocity of 4 m/s the slip achieves 10 percent. The slip gets smaller (8.5 percent) at the dip of 6° and the velocity of about 5.5 m/s.

The slip between the material and the fluid is a great disadvantage if the transportation of the slurry through a long line of flumes must be interrupted suddenly. As the water flows faster than the material, in the main-road with a length, for example, of 1000 m, the run of mine coal keeps lying in the last 200 m of the line. Starting the transportation the flow is not strong enough to accelerate the material left and blockages occur. For this kind of long flume line it is necessary to have the possibility of adding clear water automatically at the beginning of the line. Only in this manner can a sufficient slurry throughput be guaranteed.

The results obtained by measurements with the three other types of flumes are similar to that shown in this diagram. At this place it is necessary again to point to the fact that the results of these test series were only gained with flume joints without the best conditions. However, the results definitely show the advantage of the trapezoidal flume with plastic coating. At an incline of 4° it still has a very good quality for the transportation of slurry consisting of water and run of mine coal.

A German factory constructed this type of flume (figure 13). Into the trapezoidal steel flume are placed three plates of plastic called RCH 1000. This plastic material has a very low friction factor (1/3 of steel). All the plates have the same design. If the plate on the bottom after a long time is eroded to a minimum it can easily be changed for one of the plates at the flank of the same flume. So this coated flume has a three-fold life time than a normal coated flume. We expect that this type of flume, during its service life, is able to transport a volume of 3 - 4 million tons of run of mine coal.

We believe that in the mining industry the application of the plastic materials also is possible in all situations, for example, instead of wear liners consisting of steel or cast basalt.

w = water
c = coal
r = rocks

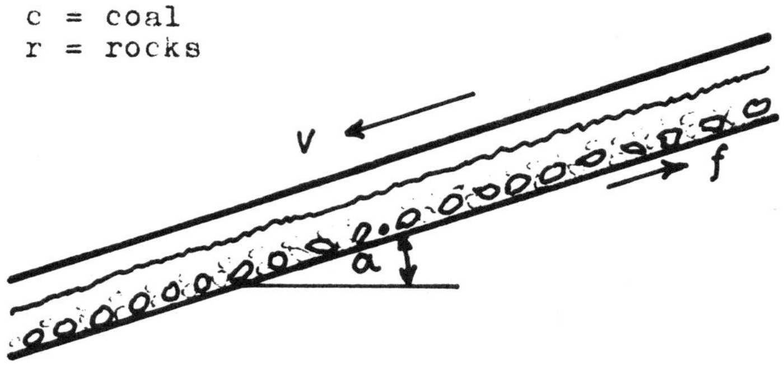


$$Q = Q_w + Q_c + Q_r$$

$$v = \frac{v_w \times Q_w - v_c \times Q_c - v_r \times Q_r}{Q}$$

Fig. 1 Physical Foundation

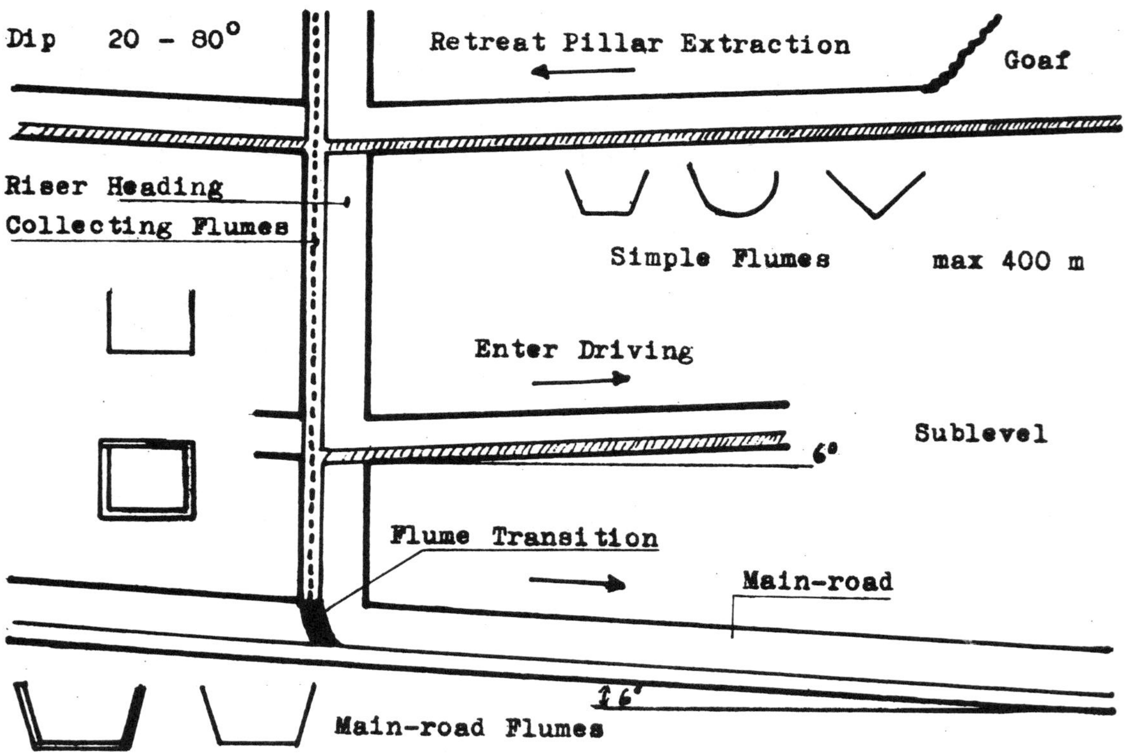


Fig. 2 Flume Systems for the hydraulic Sublevel Caving Method

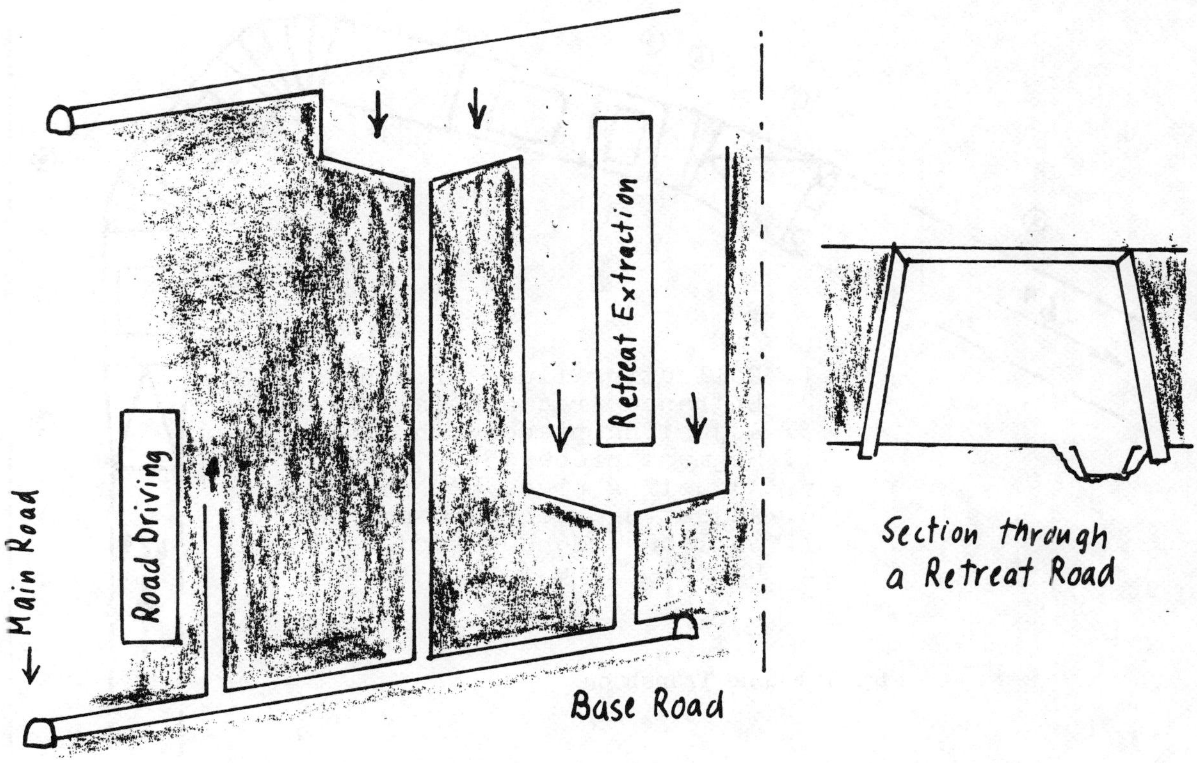


Fig. 3 Descending Pillar Extraction at an Incline less than 40°

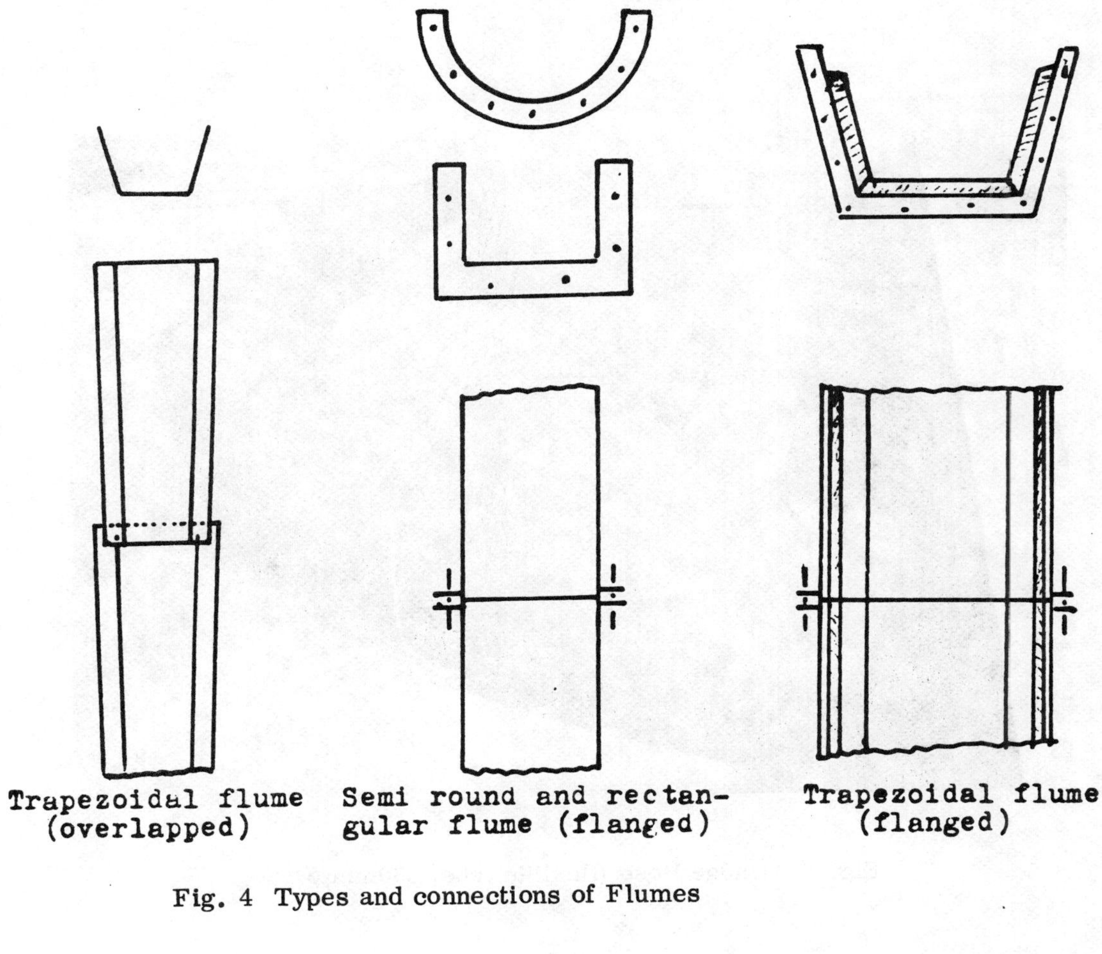


Fig. 4 Types and connections of Flumes

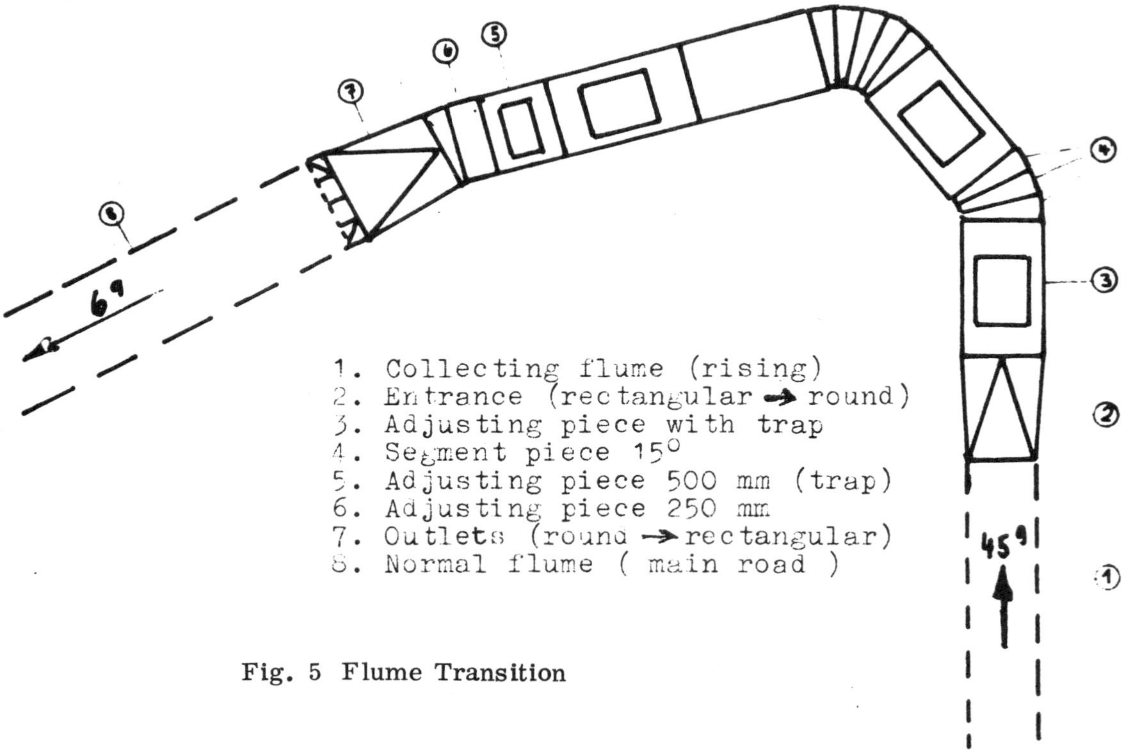


**Fig. 5 Flume Transition**

**Fig. 6 Dradge Hose (flexible tube) 650mm ø**

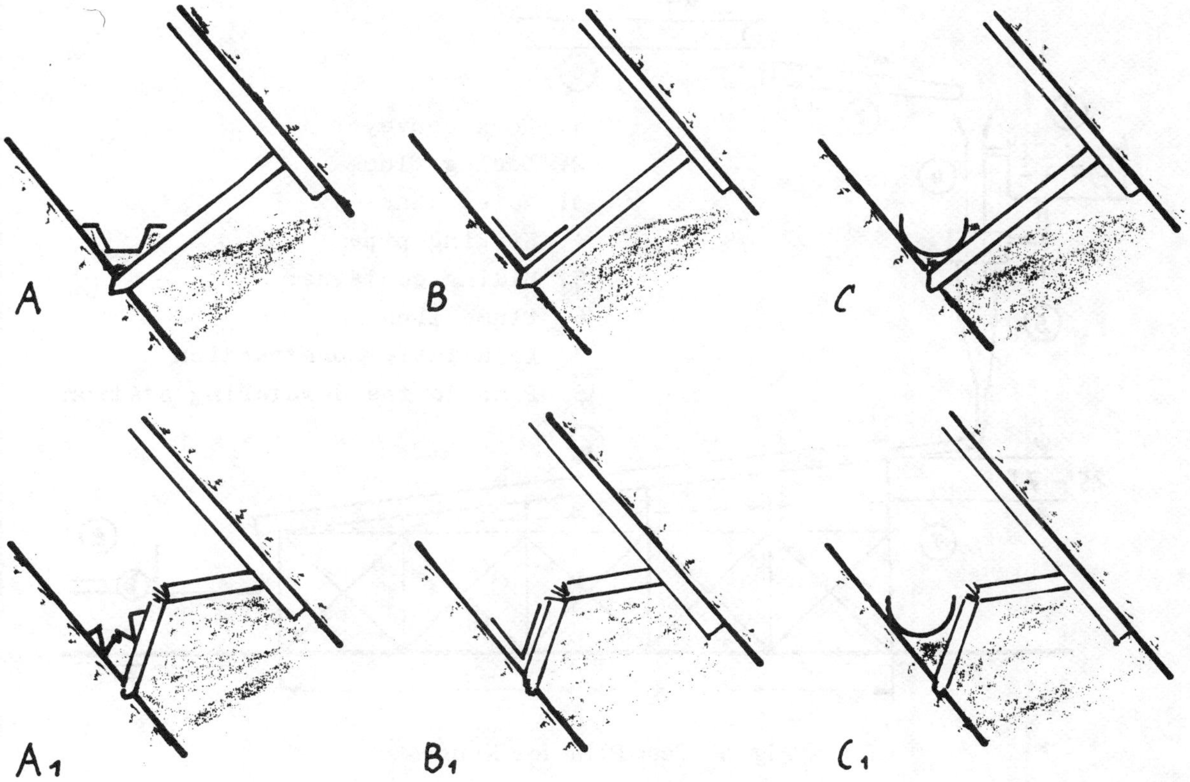


Fig. 7 Operation Characteristics of various Flume - Types by Pressure

Fig. 8 Special Flume Transition

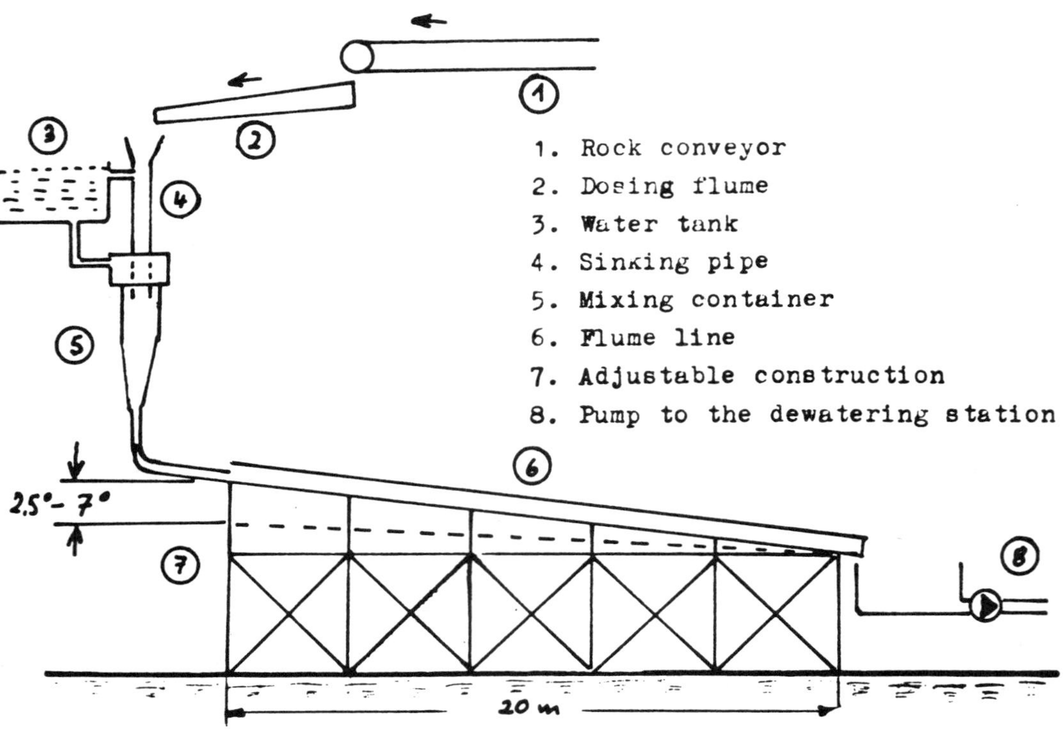


Fig. 9 Test Plant for Flumes

Fig. 10 Flume Line in the Test Plant

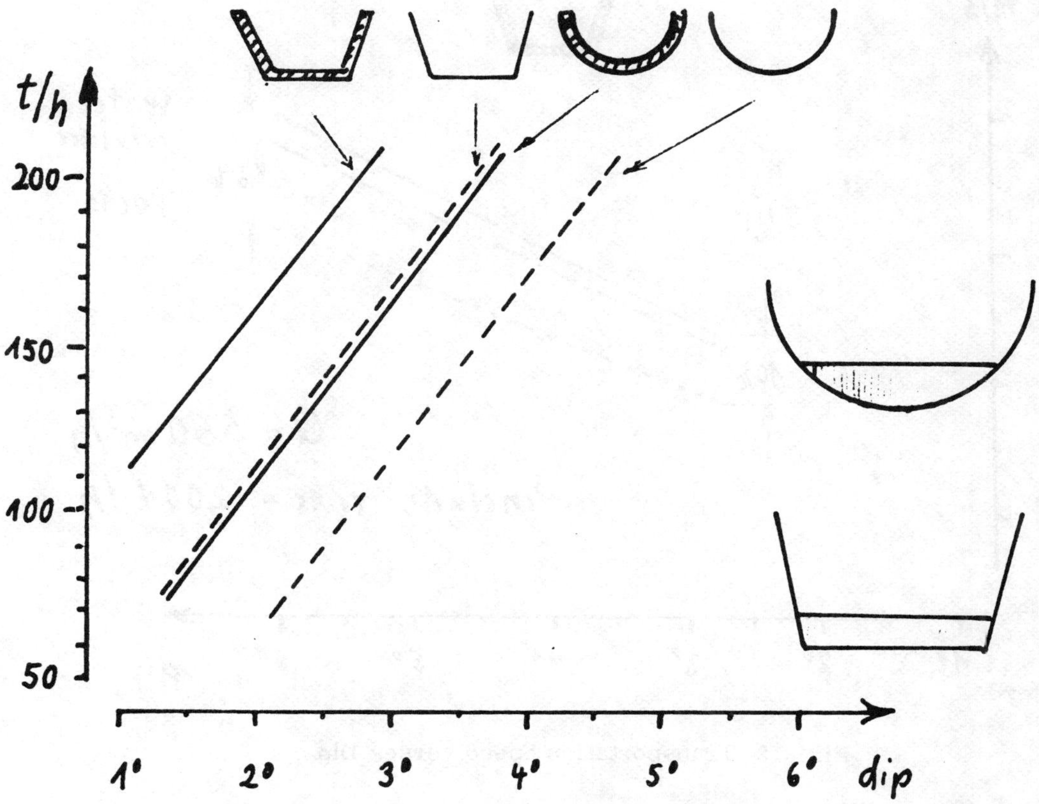


Fig. 11  Boundaries of Blockage

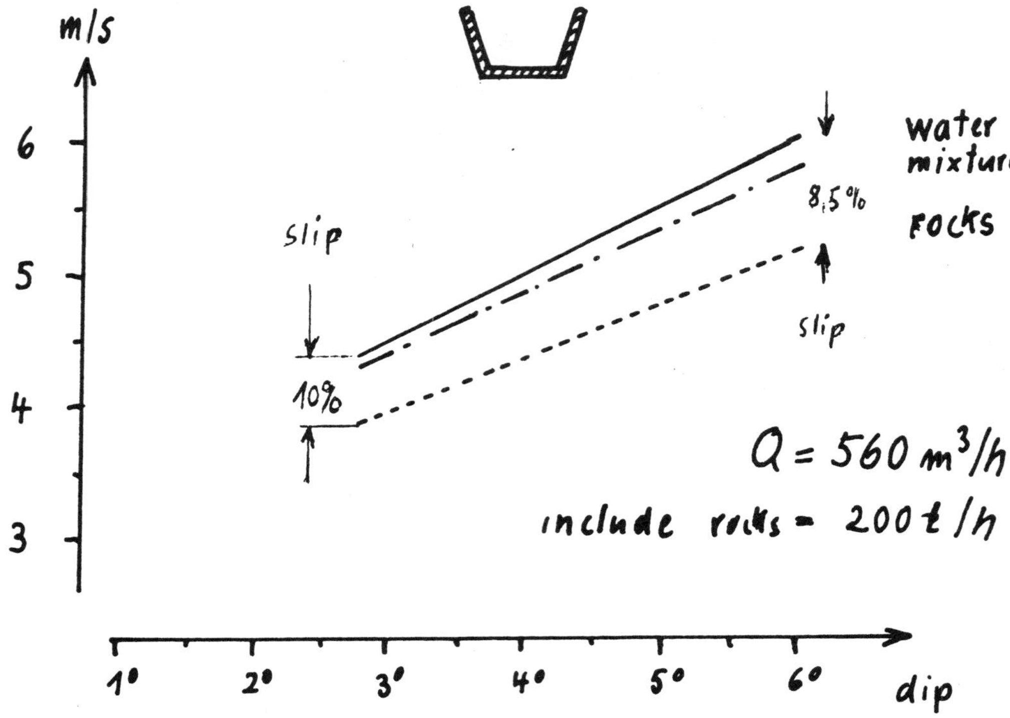


Fig. 12 Transportation Speed verses Dip

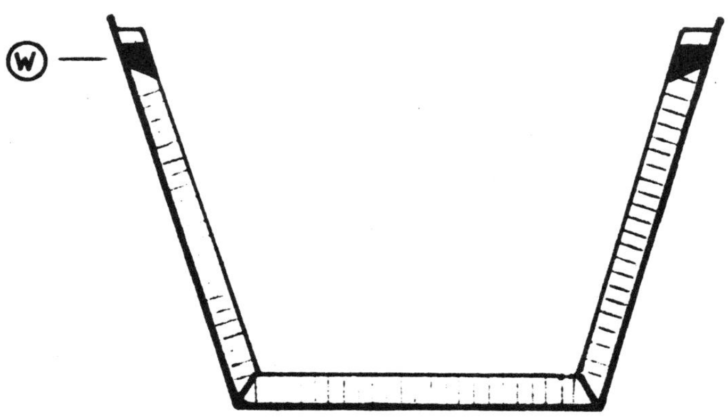


Fig. 13 Trapezoidal Flume with plastic wear liners fixed by wedges (w)

# FURTHER SELECTED BIBLIOGRAPHY

1. **A case study of a large-scale gravel pump (Palong) method of mining at Gopeng Consolidated Sdn. Bhd.**
Jackett-Simpson, R.J., Gopeng Consolidated Sdn. Bhd. In: Proc. Fourth World Conf. on Tin, (Kuala Lumpur, Malaysia: 1974), vol.2, Prospecting and Mining, Malaysia Int. Tin Council, 1974, paper 10, pp.207-221.
The paper describes the typical methods employed at a mining section of Gopeng Consolidated Sdn. Bhd., a large mining company. The general methods employed are common to most gravel pump mining operations in this locality, though particular advantage is taken of the availability of a high-pressure water supply drawn from catchment areas in the adjacent main range of hills. Catchment dams in various hill streams and main river courses have been constructed to produce a fairly constant supply of water, which is subsequently piped under pressure to the respective mining areas. Operations then utilise hydraulic monitors to break up the tin bearing-alluvium and the slurry so formed is pumped to concentration plants for separation and final upgrading of the ore. Vertical gravel pumps, electrically driven, are used to raise the slurry concentrations from level to level. Various ancillary operations are also utilised to maintain the scale of mining operations. A typical mining section at Gopeng Consolidated would cut and treat approximately 15,000 cubic yards of spoil per month, and the mining scheme at such a section is further described to indicate how these operations are performed and integrated. (A)

2. **Method of hydraulically mining coal.**
Kaiser Resources Ltd.; Mitsui Mining Co. Ltd. Brit. Pat. Spec., 1 479 619, Appl. Oct.23, 1974, Publ. Jul. 13, 1977, 8pp.
The patent describes a method of hydraulically mining coal from a seam of coal. An entry is driven into the coal seam at a slight upward angle. Monitors are then used to direct high pressure water jets at the coal seam to break up the coal. The coal water slurry thus formed flows down the sloping entry in a flume, either out of the mine if the seam is above the adit, or to an underground pumping station, where the coal may be partially dewatered if desired, and/or pumped through pipelines out of the mine. (L.G.) 11 claims; 4 figs.

3. **Marcopper's tailing disposal system – mill to sea in pipes, flumes, drop boxes.**
World Min., vol.30, no.9, Aug. 1977, pp.52-55.
25000 tons/day of tailings is being transported from a flotation mill to an undersea disposal site on the north shore of Marinduque Island in the Philippines. The design and operation of this system is described.

4. **Experience of the coal hydraulic transportation use and perspectives of its development in the enterprises of the Gidrougol' amalgamation.**
Gontov, A.E. In: Proc. Third Int. Technical Conf. on Slurry Transporation, (Las Vegas, U.S.A.: Mar. 29-31, 1978), Washington, D.C., U.S.A., Slurry Transport Assoc., 1978, pp.125-132.
The use of hydraulic transport in the coal mining industry of the U.S.S.R. is discussed. Transport of coal from the mine face is by flushing the coal cut by the continuous miner into flumes, and then by gravity flow to a central hydraulic hoist chamber. From there the coal slurry is pumped to the surface. Operating experience of such a system at two mines is discussed. (L.G)

5. **A new launder design procedure.**
Green, H.B.; Lamb, D.M.; Taylor, A.D., Bechtel Corp. Min. Engr., vol.30, no.8, Aug. 1978, pp.1210-1216.
A design procedure was developed which would take into consideration most of the known significant variables, systematise the procedure to assure consistent results and provide a rational basis for examining and utilising operating data to refine and improve the system. The procedure is described with particular reference to its mathematical basis. (A)

6. **Hydraulically mining friable carbonaceous material.**
Kaiser Resources Ltd. Brit. Pat. Spec. 1 526 739, Appl. Oct.22, 1975, Publ. Sep.27, 1978, 10pp.
The patent describes a system for hydraulically mining coal in which the monitor, (for cutting and breaking the coal) and the means of feeding and further breaking the coal and conveying it to a flume for transport as a slurry away from the face, are combined in one piece of equipment. The main monitor is mounted at the front end of the machine. The mined material is conveyed from one end of the machine through a breaking mechanism to the opposite end of the machine for discharge. An auxiliary monitor mounted part of the way along the machine is used to break up the coal and also for flushing the coal onto the conveyor system. The crushed material is transported away from the mine face as a solids/water slurry in a sloping flume. (L.G.) 15 claims; 7 figs.

7. **Water for working and transporting minerals in mining.**
Harnisch, H. Gluckauf, vol.114, no.20, Oct.19, 1978, p.426-429.
The author discusses the use of water in the hydraulic winning of coal, underground transport in flumes, hydraulic shaft raising, surface transport by slurry pipeline and the hydraulic transport of waste from the surface for stowing in level and inclined measures. Some recent applications of hydraulic transport, and the effects of hydraulic technology in mineral mining are also outlined. (D.A.L.)

8. **Hydrohandling of sweet potatoes in flumes.**
Trykowski, T.; Abrams, C.F. North Carolina Univ. Trans. Am. Soc. Agric. Engr. (Gen. Ed.), vol.22, no.5, Sep.-Oct. 1979, pp.1171-1174.
Fluming is a viable handling alternative which can be used in one farm sweet potato grading, packing and storing operations. Sweet potatoes were found easily flumable in water for water flow velocities from 30-46 m/min in practical sized channels. Allowable concentration by volume was found to be approximately 30%. (A)

9. **Vacuum-operated produce handling systems.**
Tennes, B.R.; Wolthuis, R.J.; Burton, C.L., Dept Agric. U.S. Pat.; 6 132 583, 1980, 13p. (PB80-165541).
This invention relates to the handling of fruits and vegetables during and after field harvesting. A novel flume hopper is employed in combination with a vacu-

um-operated hydraulic conveyor in field or storage handling system for efficiently transferring produce to a holding tank without inflicting damage. (A)

11. **Dealing with abrasion-corrosion. Part 2.**
Martin, J.G. Cape Breton Dev. Corp. CIM Bull., vol.73, no.823, Nov. 1980, p.149-151.
The author deals with the corrosion and erosion of steel pipes, launders and vessels in a coal preparation plant, and the required selection and installation of resistant materials throughout. Approximately 5,700 feed of various-diameter pipe and 2,000 square feet of tile work were required. The paper describes the problem, the selection and installation of a solution, and gives details and costs. (A)

10. **Further experience with horizontal and vertical hydraulic hoisting of coarse run of mine coal at Hansa Hydromine.**
Harzer, H.; Siebert, H.; Kortenbusch, W. Ruhrkohle AG In: Proc. Hydrotransport 7, 7th Int. Conf. on the Hydraulic Transport of Solids in Pipes, (Sendai, Japan: Nov.4-6, 1980), Cranfield, U.K., BHRA Fluid Engng., 1980, Paper B2, p.41-54.
At Hydrotransport 5 and 6 the transport equipment used at the Hansa Hydromine was described and the first operational experience reported on. Now, after a throughput of approx. 1.6 mill. tons of raw coal, and a 3-year period of operation more detailed particulars can be given on the technical performance and on the operating costs of the principal transport equipment. Regarded in detail within this context are the questions of classifying, batching and pressure water generation as well as the controls and instrumentation in the horizontal and vertical transport facilities. A description is furthermore given of the water clarification system, the transportation in open flumes and with the various pipelines. The operating costs incurred in 1979 are quoted in respect of the slurry flumes, the feed stations, the horizontal transport pipelines and the hoisting facilities. Important tests aimed at further optimization resulting from: combining the two feed stations; direct charging of the three-chambered pipe feeder from a clarification cone; direct charging of the three-chambered pipe feeder through the horizontal transport line are described. A report will follow in November on the technical and economic results of these tests. (A)